Praise for *Where Good Ideas Come From*

"A rapid-fire tour of 'spaces' large, small, mental, physical, and otherwise . . . *Where Good Ideas Come From* may be the ultimate distillation of his thinking on these issues. . . . One admires the intellectual athleticism of Johnson's maneuvers here."
—*The Boston Globe*

"A grand synthesis of thoughts . . . highly inventive. [Johnson] brings to the subject, however, a distinctively stimulating and enjoyable way of looking at the world, drawing not only on technology but on the history of science and medicine. He is a polymath with many stories to tell."
—*Financial Times*

"[A] rich, integrated, and often sparkling book. Mr. Johnson, who knows a thing or two about the history of science, is a first-rate storyteller."
—*The New York Times*

"A vision of innovation and ideas that is resolutely social, dynamic, and material . . . Fluidly written, entertaining, and smart without being arcane."
—*Los Angeles Times*

"A book that will stick with its readers . . . [who] are likely to find its greatest value in discoveries that arise after months or years of applying Johnson's insights to their own experiences. . . . A voyage of discovery through the history of human innovation, transporting readers from the Renaissance to the World Wide Web and beyond . . . thought-provoking in the best possible way."
—*The Dallas Morning News*

"Our standard image of the great mind is that of a solitary genius, cut off from society, finding life-changing inspiration from the stillness within. In his new exploration of creativity, Steven Johnson turns this notion on its head: good ideas come out of the thick of it. If you have not read Johnson yet, this is the time to start. Johnson enlivens his argument with stories and examples that bring personality and depth to his ideas, and make for an engaging read."
—*The Guardian*

"Johnson offers a compelling new survey of how we think, dream, and create today."
—The Daily Beast

"[D]raws on natural science, intellectual history, and twenty-first-century technology to identify the environments that are conducive to innovation . . . [Johnson's] 'long-zoom' view of fertile idea-ecosystems is engaging, informative, and, well, inspirational. . . . A magical mystery tour of the history and architecture of innovation. Whether he's illuminating the 'slow hunch' that led Wilson Greatbatch to invent the cardiac pacemaker or comparing the role of serendipity in newspapers and the Web, Johnson is always provocative and never dull."
—*The Oregonian*

WHERE GOOD IDEAS COME FROM

The
Natural History
of
Innovation

STEVEN JOHNSON

RIVERHEAD BOOKS
New York

Riverhead Books
An imprint of Penguin Random House LLC
375 Hudson Street
New York, New York 10014

Grateful acknowledgment is made for permission to reprint from Franco Moretti, *Graphs, Maps, Trees: Abstract Models for a Literary History.* Verso, 2007. Copyright © Franco Moretti 2007. All rights reserved. Reproduced by permission.

The Library of Congress has catalogued the Riverhead hardcover edition as follows:

Johnson, Steven, date.
Where good ideas come from : the natural history of innovation / Steven Johnson.
p.   cm.
ISBN 9781594487712
1. Creative thinking.   I. Title.
BF408.J56   2010   2010023481
303.48'4—dc22

First Riverhead hardcover edition: October 2010
First Riverhead trade paperback edition: October 2011
Riverhead trade paperback ISBN: 9781594485381

Printed in the United States of America
18   17

Book design by Amanda Dewey

For Peter

CONTENTS

Introduction

REEF, CITY, WEB *1*

I.

THE ADJACENT POSSIBLE *23*

II.

LIQUID NETWORKS *43*

III.

THE SLOW HUNCH *67*

IV.

SERENDIPITY *97*

V.

ERROR *129*

VI.

EXAPTATION *149*

VII.

PLATFORMS *175*

Conclusion

THE FOURTH QUADRANT *211*

Acknowledgments 247
Appendix: Chronology of Key Innovations, 1400–2000 251
Notes and Further Reading 295
Bibliography 303
Index 315

Introduction

REEF, CITY, WEB

. . . as imagination bodies forth
The forms of things unknown, the poet's pen
Turns them to shapes and gives to airy nothing
A local habitation and a name.

—SHAKESPEARE, *A Midsummer Night's Dream*, V.i.14–17

Darwin's Paradox

April 4, 1836. Over the eastern expanse of the Indian Ocean, the reliable northeast winds of monsoon season have begun to give way to the serene days of summer. On the Keeling Islands, two small atolls composed of twenty-seven coral islands six hundred miles west of Sumatra, the emerald waters are invitingly placid and warm, their hue enhanced by the brilliant white sand of disintegrated coral. On one stretch of shore usually guarded by stronger surf, the water is so calm that Charles Darwin wades out, under the vast blue sky of the tropics, to the edge of the live coral reef that rings the island.

For hours he stands and paddles among the crowded pageantry of the reef. Twenty-seven years old, seven thousand miles from London, Darwin is on the precipice, standing on an underwater peak ascending over an unfathomable sea. He is on the edge of an idea about the forces that built that peak, an idea that will prove to be the first great scientific insight of his career. And he has just begun exploring another hunch, still hazy and unformed, that will eventually lead to the intellectual summit of the nineteenth century.

Around him, the crowds of the coral ecosystem dart and shimmer. The sheer variety dazzles: butterflyfish, damselfish, parrotfish, Napoleon fish, angelfish; golden anthias feeding on plankton above the cauliflower blooms of the coral; the spikes and tentacles of sea urchins and anemones. The tableau delights Darwin's eye, but already his mind is reaching behind the surface display to a more profound mystery. In his account of the *Beagle*'s voyage, published four years later, Darwin would write: "It is excusable to grow enthusiastic over the infinite numbers of organic beings with which the sea of the tropics, so prodigal of life, teems; yet I must confess I think those naturalists who have described, in well-known words, the submarine grottoes decked with a thousand beauties, have indulged in rather exuberant language."

What lingers in the back of Darwin's mind, in the days and weeks to come, is not the beauty of the submarine grotto but rather the "infinite numbers" of organic beings. On land, the flora and fauna of the Keeling Islands are paltry at best. Among the plants, there is little but "cocoa-nut" trees, lichen, and weeds. "The list of land animals," he writes, "is even poorer than that of the plants": a handful of lizards, almost no true land birds, and those recent immigrants from European ships, rats. "The island has no domestic quadruped excepting the pig," Darwin notes with disdain.

Yet just a few feet away from this desolate habitat, in the coral reef waters, an epic diversity, rivaled only by that of the rain forests, thrives. This is a true mystery. Why should the waters at the edge of an atoll support so many different livelihoods? Extract ten thousand cubic feet of water from just about anywhere in the Indian Ocean and do a full inventory on the life you find there: the list would be about as "poor" as Darwin's account of the land animals of the Keelings. You might find a dozen fish if you were lucky. On the reef, you would be guaranteed a thousand. In Darwin's own words, stumbling across the ecosystem of a coral reef in the middle of an ocean was like encountering a swarming oasis in the middle of a desert. We now call this phenomenon Darwin's Paradox: so many different life forms, occupying such a vast array of ecological niches, inhabiting waters that are otherwise remarkably nutrient-poor. Coral reefs make up about one-tenth of one percent of the earth's surface, and yet roughly a quarter of the known species of marine life make their homes there. Darwin doesn't have those statistics available to him, standing in the lagoon in 1836, but he has seen enough of the world over the preceding four years on the *Beagle* to know there is something peculiar in the crowded waters of the reef.

The next day, Darwin ventures to the windward side of the atoll with the *Beagle*'s captain, Vice Admiral James FitzRoy, and there they watch massive waves crash against the coral's white barrier. An ordinary European spectator, accustomed to the calmer waters of the English Channel or the Mediterranean, would be naturally drawn to the impressive crest of the surf. (The breakers, Darwin observes, are almost "equal in force [to] those during a gale of wind in the temperate regions, and never cease to rage.") But Darwin has his eye on something else—not the violent surge of water but the force that resists it: the tiny organisms that have built the reef itself.

The ocean throwing its waters over the broad reef appears an invincible, all-powerful enemy; yet we see it resisted, and even conquered, by means which at first seem most weak and inefficient. It is not that the ocean spares the rock of coral; the great fragments scattered over the reef, and heaped on the beach, whence the tall cocoa-nut springs, plainly bespeak the unrelenting power of the waves . . . Yet these low, insignificant coral-islets stand and are victorious: for here another power, as an antagonist, takes part in the contest. The organic forces separate the atoms of carbonate of lime, one by one, from the foaming breakers, and unite them into a symmetrical structure. Let the hurricane tear up its thousand huge fragments; yet what will that tell against the accumulated labour of myriads of architects at work night and day, month after month?

Darwin is drawn to those minuscule architects because he believes they are the key to solving the mystery that has brought the *Beagle* to the Keeling Islands. In the Admiralty's memorandum authorizing the ship's five-year journey, one of the principal scientific directives is the investigation of atoll formation. Darwin's mentor, the brilliant geologist Charles Lyell, had recently proposed that atolls are created by undersea volcanoes that have been driven upward by powerful movements in the earth's crust. In Lyell's theory, the distinctive circular shape of an atoll emerges as coral colonies construct reefs along the circumference of the volcanic crater. Darwin's mind had been profoundly shaped by Lyell's understanding of the deep time of geological transformation, but standing on the beach, watching the breakers crash against the coral, he knows that his mentor is wrong about the origin of the atolls. It is not a story of simple geology, he realizes. It is a story about the innovative persistence of life. And as he mulls the thought, there is a hint of something else in his mind, a larger, more encompassing theory that might account for the vast scope of life's innovations. The forms of things unknown are turning, slowly, into shapes.

Days later, back on the *Beagle*, Darwin pulls out his journal and reflects on that mesmerizing clash between surf and coral. Presaging a line he would publish thirty years later in the most famous passage from *On the Origin of Species*, Darwin writes, "I can hardly explain the reason, but there is to my mind much grandeur in the view of the outer shores of these lagoon-islands." In time, the reason would come to him.

The Superlinear City

From an early age, the Swiss scientist Max Kleiber had a knack for testing the edges of convention. As an undergraduate in Zurich in the 1910s, he roamed the streets dressed in sandals and an open collar, shocking attire for the day. During his tenure in the Swiss army, he discovered that his superiors had been trading information with the Germans, despite the official Swiss position of neutrality in World War I. Appalled, he simply failed to appear at his next call-up, and was ultimately jailed for several months. By the time he had settled on a career in agricultural science, he had had enough of the restrictions of Zurich society. And so Max Kleiber charted a path that would be followed by countless sandal-wearing, nonconformist war protesters in the decades to come. He moved to California.

Kleiber set up shop at the agricultural college run by the University of California at Davis, in the heart of the fertile Central Valley. His research initially focused on cattle, measuring the impact body size had on their metabolic rates, the speed with which an organism burns through energy. Estimating metabolic rates had great practical value for the cattle industry, because it enabled farmers to predict with reasonable accuracy both how much food their livestock would require, and how much meat they would ultimately produce after slaughter. Shortly after his arrival at Davis, Kleiber stumbled across a mysterious pattern in his research, a mathematical oddity that soon brought a much more diverse array of creatures to be measured in his lab: rats, ring doves, pigeons, dogs, even humans.

Scientists and animal lovers had long observed that as life gets bigger, it slows down. Flies live for hours or days; elephants live for half-centuries. The hearts of birds and small mammals pump blood much faster than those of giraffes and blue whales. But the relationship between size and speed didn't seem to be a linear one. A horse might be five hundred times heavier than a rabbit, yet its pulse certainly wasn't five hundred times slower than the rabbit's. After a formidable series of measurements in his Davis lab, Kleiber discovered that this scaling phenomenon stuck to an unvarying mathematical script called "negative quarter-power scaling." If you plotted mass versus metabolism on a logarithmic grid, the result was a perfectly straight line that led from rats and pigeons all the way up to bulls and hippopotami.

Physicists were used to discovering beautiful equations like this lurking in the phenomena they studied, but mathematical elegance was a rarity in the comparatively messy world of biology. But the more species Kleiber and his peers analyzed, the clearer the equation became: metabolism scales to mass to the negative quarter power. The math is simple enough: you take the square root of 1,000, which is (approximately) 31, and then take the square root of 31, which is (again, approximately) 5.5. This means that a cow, which is roughly a thousand times heavier than a woodchuck, will, on average, live 5.5 times longer, and have a heart rate that is 5.5 times slower than the woodchuck's. As the science writer George Johnson once observed, one lovely consequence of Kleiber's law is that the number of heartbeats per lifetime tends to be stable from species to species. Bigger animals just take longer to use up their quota.

Over the ensuing decades, Kleiber's law was extended down to the microscopic scale of bacteria and cell metabolism; even plants were found to obey negative quarter-power scaling in their patterns of growth. Wherever life appeared, whenever an organism had to figure out a way to consume and distribute energy through a body, negative quarter-power scaling governed the patterns of its development.

Several years ago, the theoretical physicist Geoffrey West decided to investigate whether Kleiber's law applied to one of life's largest creations: the superorganisms of human-built cities. Did the "metabolism" of urban life slow down as cities grew in size? Was there an underlying pattern to the growth and pace of life of metropolitan systems? Working out of the legendary Santa Fe Institute, where he served as president until 2009, West assembled an international team of researchers and advisers to collect data on dozens of cities around the world, measuring everything from crime to household electrical consumption, from new patents to gasoline sales.

When they finally crunched the numbers, West and his team were delighted to discover that Kleiber's negative quarter-power scaling governed the energy and transportation growth of city living. The number of gasoline stations, gasoline sales, road surface area, the length of electrical cables: all these factors follow the exact same power law that governs the speed with which energy is expended in biological organisms. If an elephant was just a scaled-up mouse, then, from an energy perspective, a city was just a scaled-up elephant.

But the most fascinating discovery in West's research came from the data that *didn't* turn out to obey Kleiber's law. West and his team discovered another power law lurking in their immense database of urban statistics. Every datapoint that involved creativity and innovation—patents, R&D budgets, "supercreative" professions, inventors—also followed a quarter-power law, in a way that was every bit as predictable as Kleiber's law. But there was one fundamental difference: the quarter-power law governing innovation was *positive*, not negative. A city that was ten times larger than its neighbor wasn't ten times more innovative; it was *seventeen* times more innovative. A metropolis fifty times bigger than a town was 130 times more innovative.

Kleiber's law proved that as life gets bigger, it slows down. But West's model demonstrated one crucial way in which human-built cities broke from the patterns of biological life: as cities get bigger, they generate ideas at a faster clip. This is what we call "superlinear scaling": if creativity scaled with size in a straight, linear fashion, you would of course find more patents and inventions in a larger city, but the number of patents and inventions per capita would be stable. West's power laws suggested something far more provocative: that despite all the noise and crowding and distraction, the average resident of a metropolis with a population of five million people was almost *three times* more creative than the average resident of a town of a hundred thousand. "Great cities are not like towns only larger," Jane Jacobs wrote nearly fifty years ago. West's positive quarter-power law gave that insight a mathematical foundation. Something about the environment of a big city was making its residents significantly more innovative than residents of smaller towns. But what was it?

The 10/10 Rule

The first national broadcast of a color television program took place on January 1, 1954, when NBC aired an hour-long telecast of the Tournament of Roses parade, and distributed it to twenty-two cities across the country. For those lucky enough to see the program, the effect of a moving color image on a small screen seems to have been mesmerizing. The *New York Times*, in typical language, called it a "veritable bevy of hues and depth." "To concentrate so much color information within the frame of a small screen," the *Times* wrote, "would be difficult for even the most gifted artist doing a 'still' painting. To do it with constantly moving pictures seemed pure wizardry." Alas, the Rose Parade "broadcast" turned out to be not all that broad, given that it was visible only on prototype televisions in RCA showrooms. Color programming would not become standard on prime-time shows until the late 1960s. After the advent of color, the basic conventions that defined the television image would go unchanged for decades. The delivery mechanisms began to diversify with the introduction of VCRs and cable in the late 1970s. But the image remained the same.

In the mid-1980s, a number of influential media and technol-

ogy executives, along with a few visionary politicians, had the eminently good idea that it was time to upgrade the video quality of
broadcast television. Speeches were delivered, committees formed,
experimental prototypes built, but it wasn't until July 23, 1996, that
a Raleigh, North Carolina, CBS affiliate initiated the first public
transmission of an HDTV signal. Like the Tournament of Roses
footage, though, there were no ordinary consumers with sets capable
of displaying its "wizardry."[1] A handful of broadcasters began trans-
mitting HDTV signals in 1999, but HD television didn't become a

---

1. The convoluted history of HDTV's origins could be the subject of an entire book, but the
condensed version goes something like this: in the early 1980s the Japanese public broadcasting
company NHK gave a series of demonstrations of a prototype high-definition television platform to members of the U.S. Congress and other government officials. This was at the height
of American fears about Japan's economic ascendancy, a time when Sony televisions were already outselling venerable American brands like RCA and Zenith. The idea that the Japanese
might introduce a higher-quality image to the U.S. market posed a threat both to American
consumer electronics companies and, as then-senator Al Gore pointed out after watching the
NHK demo, to the semiconductor companies that would make the chips for all those new television boxes. Within a matter of months, the Federal Communications Commission formally
decided to investigate the possibility of improving the picture quality of broadcast and cable
TV. All the forces were aligned for the next major step forward in the television medium. Ronald Reagan, always one to grasp the transformative possibilities of television, even called the
development of a U.S. HDTV standard a matter of "national interest."

But what followed in the subsequent years was less of a Great Leap Forward and more
of an endless, serpentine crawl. First, the FCC appointed a committee—the Advisory Committee on Advanced Television Service (ACATS)—that solicited and reviewed twenty-three different proposals over the next year, eventually winnowing them down to six different systems, each
using a unique scheme to convey higher-definition sound and image. Some were analog, others
digital. Some were backward compatible with the current systems; others would require the
consumer to upgrade to new equipment. For five years, the sponsor organizations enhanced and
tested their various platforms, at a cost of hundreds of millions in research-and-development
dollars. The whole process was supposed to come to a conclusion in 1993, when ACATS was
scheduled to run a series of final tests and pick a winner, but the final tests turned out to be a
preamble: the only thing the committee agreed on was that digital was preferable to analog,
which reduced the field slightly. The remaining contenders all had enough flaws individually
to keep the committee from anointing a new heir apparent, and so the ACATS group proposed
that the remaining candidates collaborate on a single standard. This group—called the Grand
Alliance—reached agreement on specifications for digital high-definition video and audio in
1995, which the FCC embraced the following year.

mainstream consumer phenomenon for another five years. Even after the FCC mandated that all television stations cease broadcasting the old analog standard on June 12, 2009, more than 10 percent of U.S. households had televisions that went dark that day.

It is one of the great truisms of our time that we live in an age of technological *acceleration*; the new paradigms keep rolling in, and the intervals between them keep shortening. This acceleration reflects not only the flood of new products, but also our growing willingness to embrace these strange new devices, and put them to use. The waves roll in at ever-increasing frequencies, and more and more of us are becoming trained surfers, paddling out to meet them the second they start to crest. But the HDTV story suggests that this acceleration is hardly a universal law. If you measure how quickly a new technology progresses from an original idea to mass adoption, then it turns out that HDTV was traveling at the exact same speed that color television had traveled four decades earlier. It took ten years for color TV to go from the fringes to the mainstream; two generations later, it took HDTV just as long to achieve mass success.

In fact, if you look at the entirety of the twentieth century, the most important developments in mass, one-to-many communications clock in at the same social innovation rate with an eerie regularity. Call it the 10/10 rule: a decade to build the new platform, and a decade for it to find a mass audience. The technology standard of amplitude-modulated radio—what we now call AM radio—evolved in the first decade of the twentieth century. The first commercial AM station began broadcasting in 1920, but it wasn't until the late 1920s that radios became a fixture in American households. Sony inaugurated research into the first consumer videocassette recorder in 1969, but didn't ship its first Betamax for another seven years, and VCRs didn't become a household necessity until the mid-eighties. The DVD player didn't statistically replace the VCR in American households until 2006, nine years after the first players went on the market. Cell phones, personal computers, GPS navigation devices—all took a similar time frame to go from innovation to mass adoption.

Consider, as an alternate scenario, the story of Chad Hurley, Steve Chen, and Jawed Karim, three former employees of the online payment site PayPal, who decided in early 2005 that the Web was ripe for an upgrade in the way it handled video and sound. Video, of course, was not native to the Web, which had begun its life fifteen years before as a platform for academics to share hypertext documents. But over the years, video clips had begun to trickle their way online, thanks to new video standards that emerged, such as QuickTime, Flash, or Windows Media Player. But the mechanisms that allowed people to upload and share their own videos were too challenging for most ordinary users. So Hurley, Chen, and Karim cobbled together a rough beta for a service that would correct these deficiencies, raised less than $10 million in venture capital, hired about two dozen people, and launched YouTube, a website that utterly transformed the way video information is shared online. Within sixteen months of the company's founding, the service was streaming more than 30 million videos a day. Within two years, YouTube was one of the top-ten most visited sites on the Web. Before Hurley, Chen, and Karim hit upon their idea for a start-up, video on the Web was as common as subtitles on television. The Web was about doing things with text, and uploading the occasional photo. YouTube brought Web video into the mainstream.

Now compare the way these two ideas—HDTV and YouTube—

changed the basic rules of engagement for their respective platforms. Going from analog television to HDTV is a change in degree,
not in kind: there are more pixels; the sound is more immersive; the
colors are sharper. But consumers watch HDTV the exact same way
they watched old-fashioned analog TV. They choose a channel, and
sit back and watch. YouTube, on the other hand, radically altered the
basic rules of the medium. For starters, it made watching video on
the Web a mass phenomenon. But with YouTube you weren't limited
to sitting and watching a show, television-style; you could also upload your own clips, recommend or rate other clips, get into a conversation about them. With just a few easy keystrokes, you could take
a clip running on someone else's site, and drop a copy of it onto your
own site. The technology allowed ordinary enthusiasts to effectively
program their own private television networks, stitching together
video clips from all across the planet.

Some will say that this is merely a matter of software, which is
intrinsically more adaptable than hardware like televisions or cellular phones. But before the Web became mainstream in the mid-
1990s, the pace of software innovation followed the exact same
10/10 pattern of development that we saw in the spread of other
twentieth-century technologies. The graphical user interface, for instance, dates back to a famous technology demo given by pioneering
computer scientist Doug Engelbart in 1968. During the 1970s, many
of its core elements—like the now ubiquitous desktop metaphor—
were developed by researchers at Xerox-PARC. But the first commercial product with a fully realized graphical user interface didn't
ship until 1981, in the form of the Xerox Star workstation, followed
by the Macintosh in 1984, the first graphical user interface to reach
a mainstream, if niche, audience. But it wasn't until the release of

Windows 3.0 in 1990—almost exactly ten years after the Xerox Star
hit the market—that graphical user interfaces became the norm.
The same pattern occurs in the developmental history of other software genres, such as word processors, spreadsheets, or e-mail clients.
They were all built out of bits, not atoms, but they took just as long
to go from idea to mass success as HDTV did.

There are many ways to measure innovation, but perhaps the
most elemental yardstick, at least where technology is concerned,
revolves around the *job* that the technology in question lets you do.
All other things being equal, a breakthrough that lets you execute
two jobs that were impossible before is twice as innovative as a
breakthrough that lets you do only one new thing. By that measure,
YouTube was significantly more innovative than HDTV, despite the
fact that HDTV was a more complicated technical problem. You-
Tube let you publish, share, rate, discuss, and watch video more
efficiently than ever before. HDTV let you watch more pixels than
ever before. But even with all those extra layers of innovation, You-
Tube went from idea to mass adoption in less than two years. Something about the Web environment had enabled Hurley, Chen, and
Karim to unleash a good idea on the world with astonishing speed.
They took the 10/10 rule and made it 1/1.

This is a book about the space of innovation. Some environments squelch new ideas; some environments seem to breed
them effortlessly. The city and the Web have been such engines of
innovation because, for complicated historical reasons, they are both
environments that are powerfully suited for the creation, diffusion,
and adoption of good ideas. Neither environment is perfect, by any

means. (Think of crime rates in big cities, or the explosion of spam online.) But both the city and the Web possess an undeniable track record at generating innovation.[2] In the same way, the "myriad tiny architects" of Darwin's coral reef create an environment where biological innovation can flourish. If we want to understand where good ideas come from, we have to put them in context. Darwin's world-changing idea unfolded inside his brain, but think of all the environments and tools he needed to piece it together: a ship, an archipelago, a notebook, a library, a coral reef. Our thought shapes the spaces we inhabit, and our spaces return the favor. The argument of this book is that a series of shared properties and patterns recur again and again in unusually fertile environments. I have distilled them down into seven patterns, each one occupying a separate chapter. The more we embrace these patterns—in our private work habits and hobbies, in our office environments, in the design of new software tools—the better we will be at tapping our extraordinary capacity for innovative thinking.[3]

---

2. This fact, ironically, may be related to some of their blemishes. It may be that the criminals and spammers thrive in these spaces because they, too, are able to be more innovative at their trades.
3. Sections of the argument that follows will be familiar to anyone who has spent the last decade or two exploring the new possibility spaces of the Web. I last wrote about the Web in book form ten years ago; since that time, a marvelous community of entrepreneur theorists has materialized, capable of pushing the boundaries of the medium, and at the same time reflecting on what those advances might mean. We have, all of us, seen firsthand how innovative a space the Web can be, and we have assembled a great deal of local knowledge about the forces that make that innovation possible. In assembling the seven patterns of innovation, I have tried to organize that knowledge into productive categories, and I hope I have provided a few insights into how the Web works that will surprise the natives. But even the most devoted crowd-sourcing, microblogging Wikipedia-head has doubts about how portable the Web experience is to real-world innovation environments. Just because the patterns work for Google doesn't mean that they are relevant for an understaffed nonprofit, or auto-parts manufacturer, or city government. And so one way to think about the pages that follow is as an argument that the particular magic that we have seen on the Web has a long history that predates the Web and can be reproduced in other environments.

These patterns turn out to have a long history, much older than most of the systems that we conventionally associate with innovation. This history is particularly rich because it is not exclusively limited to human creations like the Internet or the metropolis. The amplification and adoption of useful innovation exist throughout *natural* history as well. Coral reefs are sometimes called the "cities of the sea," and part of the argument of this book is that we need to take the metaphor seriously: the reef ecosystem is so innovative in its exploitation of those nutrient-poor waters because it shares some defining characteristics with actual cities. In the language of complexity theory, these patterns of innovation and creativity are fractal: they reappear in recognizable form as you zoom in and out, from molecule to neuron to pixel to sidewalk. Whether you're looking at the original innovations of carbon-based life, or the explosion of new software tools on the Web, the same shapes keep turning up. When life gets creative, it has a tendency to gravitate toward certain recurring patterns, whether those patterns are emergent and self-organizing, or whether they are deliberately crafted by human agents.

It may seem odd to talk about such different regions of experience as though they were interchangeable. But in fact, we are constantly making equivalent conceptual leaps from biology to culture without blinking. It is not a figure of speech to say that the pattern of "competition"—a term often associated with innovation—plays a critical role in the behavior of marketplaces, in the interaction between a swarm of sperm cells and an egg, and in the ecosystem-scale battle between organisms for finite energy sources. We are not using a metaphor of economic competition to describe the struggles of those sperm cells: the meaning of the word "competition" is

wide (or perhaps deep) enough to encompass sperm cells *and* corporations. The same principle applies to the seven patterns I have
assembled here.

Traveling across these different environments and scales is not
merely intellectual tourism. Science long ago realized that we can
understand something better by studying its behavior in different
contexts. When we want to answer a question like "Why has the
Web been so innovative?" we naturally invoke thoughts of its creators, and the workspaces, organizations, and information networks
they used in building it. But it turns out that we can answer the
question more comprehensively if we draw analogies to patterns of
innovation that we see in ecosystems like Darwin's coral reef, or in
the structure of the human brain. We have no shortage of theories
to instruct us how to make our organizations more creative, or explain why tropical rain forests engineer so much molecular diversity. What we lack is a unified theory that describes the common
attributes shared by all those innovation systems. Why is a coral
reef such an engine of biological innovation? Why do cities have
such an extensive history of idea creation? Why was Darwin able to
hit upon a theory that so many brilliant contemporaries of his
missed? No doubt there are partial answers to these questions that
are unique to each situation, and each scale: the ecological history
of the reef; the sociology of urban life; the intellectual biography of
a scientist. But the argument of this book is that there are other, more
interesting answers that are applicable to all three situations, and that
by approaching the problem in this fractal, cross-disciplinary way,
new insights become visible. Watching the ideas spark on these
different scales reveals patterns that single-scale observations easily
miss or undervalue.

I call that vantage point the *long zoom*. It can be imagined as a kind of hourglass:

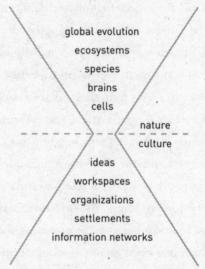

As you descend toward the center of the glass, the biological scales contract: from the global, deep time of evolution to the microscopic exchanges of neurons or DNA. At the center of the glass, the perspective shifts from nature to culture, and the scales widen: from individual thoughts and private workspaces to immense cities and global information networks. When we look at the history of innovation from the vantage point of the long zoom, what we find is that unusually generative environments display similar patterns of creativity at multiple scales simultaneously. You can't explain the biodiversity of the coral reef by simply studying the genetics of the coral itself. The reef generates and sustains so many different forms of life because of patterns that recur on the scales of cells, organisms, and the wider ecosystem itself. The sources of innovation in the city and the Web are equally fractal. In this sense, seeing the problem of innovation from the long-zoom perspective does not just give us new metaphors. It gives us new *facts*.

The pattern of "competition" is an excellent case in point. Every economics textbook will tell you that competition between rival firms leads to innovation in their products and services. But when you look at innovation from the long-zoom perspective, competition turns out to be less central to the history of good ideas than we generally think. Analyzing innovation on the scale of individuals and organizations—as the standard textbooks do—distorts our view. It creates a picture of innovation that overstates the role of proprietary research and "survival of the fittest" competition. The long-zoom approach lets us see that openness and connectivity may, in the end, be more valuable to innovation than purely competitive mechanisms. Those patterns of innovation deserve recognition—in part because it's intrinsically important to understand why good ideas emerge historically, and in part because by embracing these patterns we can build environments that do a better job of nurturing good ideas, whether those environments are schools, governments, software platforms, poetry seminars, or social movements. We can think more creatively if we open our minds to the many connected environments that make creativity possible.

The academic literature on innovation and creativity is rich with subtle distinctions between innovations and inventions, between different modes of creativity: artistic, scientific, technological. I have deliberately chosen the broadest possible phrasing—good ideas—to suggest the cross-disciplinary vantage point I am trying to occupy. The good ideas in this survey range from software platforms to musical genres to scientific paradigms to new models for government. My premise is that there is as much value to be found in seeking the common properties across all these varied forms of innovation and creativity as there is value to be found in documenting the differences between them. The poet and the engineer (and the coral reef) may seem a million miles apart in their particular forms of expertise, but when they bring good ideas into the world, similar patterns of development and collaboration shape that process.

If there is a single maxim that runs through this book's arguments, it is that we are often better served by *connecting* ideas than we are by protecting them. Like the free market itself, the case for restricting the flow of innovation has long been buttressed by appeals to the "natural" order of things. But the truth is, when one looks at innovation in nature and in culture, environments that build walls around good ideas tend to be less innovative in the long run than more open-ended environments. Good ideas may not want to be free, but they do want to connect, fuse, recombine. They want to reinvent themselves by crossing conceptual borders. They want to complete each other as much as they want to compete.

I.

THE ADJACENT
POSSIBLE

Sometime in the late 1870s, a Parisian obstetrician named Stephane Tarnier took a day off from his work at Maternité de Paris, the lying-in hospital for the city's poor women, and paid a visit to the nearby Paris Zoo. Wandering past the elephants and reptiles and classical gardens of the zoo's home inside the Jardin des Plantes, Tarnier stumbled across an exhibit of chicken incubators. Seeing the hatchlings totter about in the incubator's warm enclosure triggered an association in his head, and before long he had hired Odile Martin, the zoo's poultry raiser, to construct a device that would perform a similar function for human newborns. By modern standards, infant mortality was staggeringly high in the late nineteenth century, even in a city as sophisticated as Paris. One in five babies died before learning to crawl, and the odds were far worse for premature babies born with low birth weights. Tarnier knew that temperature regulation was critical for keeping these infants alive, and he knew that the French medical establishment had a deep-seated obsession with statistics. And so as soon as his newborn incubator had been installed at Maternité, the fragile infants warmed by hot water bottles below the wooden boxes, Tarnier embarked on a quick study of five hundred babies. The results shocked the Parisian medical establishment: while 66 percent of low-weight babies died within weeks of birth, only 38 percent died if they were housed in Tarnier's incubating box. You could effectively halve the mortality rate for premature babies simply by treating them like hatchlings in a zoo.

Tarnier's incubator was not the first device employed for warming newborns, and the contraption he built with Martin would be improved upon significantly in the subsequent decades. But Tarnier's statistical analysis gave newborn incubation the push that it needed: within a few years, the Paris municipal board required that incubators be installed in all the city's maternity hospitals. In 1896, an enterprising physician named Alexandre Lion set up a display of incubators—with live newborns—at the Berlin Exposition. Dubbed the *Kinderbrutenstalt*, or "child hatchery," Lion's exhibit turned out to be the sleeper hit of the exposition, and launched a bizarre tradition of incubator sideshows that persisted well into the twentieth century. (Coney Island had a permanent baby incubator show until the early 1940s.) Modern incubators, supplemented with high-oxygen therapy and other advances, became standard equipment in all American hospitals after the end of World War II, triggering a spectacular 75 percent decline in infant mortality rates between 1950 and 1998. Because incubators focus exclusively on the beginning of life, their benefit to public health—measured by the sheer number of extra years they provide—rivals any medical advance of the twentieth century. Radiation therapy or a double bypass might give you another decade or two, but an incubator gives you an entire lifetime.

In the developing world, however, the infant mortality story remains bleak. Whereas infant deaths are below ten per thousand births throughout Europe and the United States, over a hundred infants die per thousand in countries like Liberia and Ethiopia, many of them premature babies that would have survived with access to incubators. But modern incubators are complex, expensive things. A standard incubator in an American hospital might cost more than $40,000. But the expense is arguably the smaller hurdle to overcome. Complex equipment breaks, and when it breaks you need the technical expertise to fix it, and you need replacement parts. In the year that followed the 2004 Indian Ocean tsunami, the Indonesian city of Meulaboh received eight incubators from a range of international relief organizations. By late 2008, when an MIT professor named Timothy Prestero visited the hospital, all eight were out of order, the victims of power surges and tropical humidity, along with the hospital staff's inability to read the English repair manual. The Meulaboh incubators were a representative sample: some studies suggest that as much as 95 percent of medical technology donated to developing countries breaks within the first five years of use.

Prestero had a vested interest in those broken incubators, because the organization he founded, Design that Matters, had been working for several years on a new scheme for a more reliable, and less expensive, incubator, one that recognized complex medical technology was likely to have a very different tenure in a developing world context than it would in an American or European hospital. Designing an incubator for a developing country wasn't just a matter of creating something that worked; it was also a matter of designing something that would break in a non-catastrophic way. You couldn't guarantee a steady supply of spare parts, or trained repair technicians. So instead, Prestero and his team decided to build an incubator out of parts that were already abundant in the developing world. The idea had originated with a Boston doctor named Jonathan Rosen, who had observed that even the smaller towns of the developing world seemed to be able to keep automobiles in working order. The towns might have lacked air conditioning and laptops and cable television, but they managed to keep their Toyota 4Runners on the road. So Rosen approached Prestero with an idea: What if you made an incubator out of automobile parts?

Three years after Rosen suggested the idea, the Design that Matters team introduced a prototype device called the NeoNurture. From the outside, it looked like a streamlined modern incubator, but its guts were automotive. Sealed-beam headlights supplied the crucial warmth; dashboard fans provided filtered air circulation; door chimes sounded alarms. You could power the device via an adapted cigarette lighter, or a standard-issue motorcycle battery. Building the NeoNurture out of car parts was doubly efficient, because it tapped both the local supply of parts themselves and the local knowledge of automobile repair. These were both abundant resources in the developing world context, as Rosen liked to say. You didn't have to be a trained medical technician to fix the NeoNurture; you didn't even have to read the manual. You just needed to know how to replace a broken headlight.

Good ideas are like the NeoNurture device. They are, inevitably, constrained by the parts and skills that surround them. We have a natural tendency to romanticize breakthrough innovations, imagining momentous ideas transcending their surroundings, a gifted mind somehow seeing over the detritus of old ideas and ossified tradition. But ideas are works of bricolage; they're built out of that detritus. We take the ideas we've inherited or that we've stumbled across, and we jigger them together into some new shape. We like to think of our ideas as $40,000 incubators, shipped direct from the factory, but in reality they've been cobbled together with spare parts that happened to be sitting in the garage.

Before his untimely death in 2002, the evolutionary biologist Stephen Jay Gould maintained an odd collection of footware that he had purchased during his travels through the developing world, in open-air markets in Quito, Nairobi, and Delhi. They were sandals made from recycled automobile tires. As a fashion statement, they may not have amounted to much, but Gould treasured his tire sandals as a testimony to "human ingenuity." But he also saw them as a metaphor for the patterns of innovation in the biological world. Nature's innovations, too, rely on spare parts. Evolution advances by taking available resources and cobbling them together to create new uses. The evolutionary theorist François Jacob captured this in his concept of evolution as a "tinkerer," not an engineer; our bodies are also works of bricolage, old parts strung together to form something radically new. "The tires-to-sandals principle works at all scales and times," Gould wrote, "permitting odd and unpredictable initiatives at any moment—to make nature as inventive as the cleverest person who ever pondered the potential of a junkyard in Nairobi."

You can see this process at work in the primordial innovation of life itself. We do not yet have scientific consensus on the specifics of life's origins. Some believe life originated in the boiling, metallic vents of undersea volcanoes; others suspect the open oceans; others point to the tidal ponds where Darwin believed life first took hold.

Many respected scientists think that life may have arrived from outer space, embedded in a meteor. But we have a much clearer picture of the composition of earth's atmosphere before life emerged, thanks to a field known as prebiotic chemistry. The lifeless earth was dominated by a handful of basic molecules: ammonia, methane, water, carbon dioxide, a smattering of amino acids, and other simple organic compounds. Each of these molecules was capable of a finite series of transformations and exchanges with other molecules in the primordial soup: methane and oxygen recombining to form formaldehyde and water, for instance.

Think of all those initial molecules, and then imagine all the potential new combinations that they could form spontaneously, simply by colliding with each other (or perhaps prodded along by the extra energy of a propitious lightning strike). If you could play God and trigger all those combinations, you would end up with most of the building blocks of life: the proteins that form the boundaries of cells; sugar molecules crucial to the nucleic acids of our DNA. But you would not be able to trigger chemical reactions that would build a mosquito, or a sunflower, or a human brain. Formaldehyde is a first-order combination: you can create it directly from the molecules in the primordial soup. The atomic elements that make up a sunflower are the very same ones available on earth before the emergence of life, but you can't spontaneously create a sunflower in that environment, because it relies on a whole series of subsequent innovations that wouldn't evolve on earth for billions of years: chloroplasts to capture the sun's energy, vascular tissues to circulate resources through the plant, DNA molecules to pass on sunflower-building instructions to the next generation.

The scientist Stuart Kauffman has a suggestive name for the

set of all those first-order combinations: "the adjacent possible." The phrase captures both the limits and the creative potential of change and innovation. In the case of prebiotic chemistry, the adjacent possible defines all those molecular reactions that were directly achievable in the primordial soup. Sunflowers and mosquitoes and brains exist outside that circle of possibility. The adjacent possible is a kind of shadow future, hovering on the edges of the present state of things, a map of all the ways in which the present can reinvent itself. Yet is it not an infinite space, or a totally open playing field. The number of potential first-order reactions is vast, but it is a finite number, and it excludes most of the forms that now populate the biosphere. What the adjacent possible tells us is that at any moment the world is capable of extraordinary change, but only *certain* changes can happen.

The strange and beautiful truth about the adjacent possible is that its boundaries grow as you explore those boundaries. Each new combination ushers new combinations into the adjacent possible. Think of it as a house that magically expands with each door you open. You begin in a room with four doors, each leading to a new room that you haven't visited yet. Those four rooms are the adjacent possible. But once you open one of those doors and stroll into that room, three new doors appear, each leading to a brand-new room that you couldn't have reached from your original starting point. Keep opening new doors and eventually you'll have built a palace.

Basic fatty acids will naturally self-organize into spheres lined with a dual layer of molecules, very similar to the membranes that define the boundaries of modern cells. Once the fatty acids combine to form those bounded spheres, a new wing of the adjacent possible opens up, because those molecules implicitly create a fundamental division between the inside and outside of the sphere. This division is the very essence of a cell. Once you have an "inside," you can put things there: food, organelles, genetic code. Small molecules can pass through the membrane and then combine with other molecules to form larger entities too big to escape back through the boundaries of the proto-cell. When the first fatty acids spontaneously formed those dual-layered membranes, they opened a door into the adjacent possible that would ultimately lead to nucleotide-based genetic code, and the power plants of the chloroplasts and mitochondria—the primary "inhabitants" of all modern cells.

The same pattern appears again and again throughout the evolution of life. Indeed, one way to think about the path of evolution is as a continual exploration of the adjacent possible. When dinosaurs such as the velociraptor evolved a new bone called the semi-lunate carpal (the name comes from its half-moon shape), it enabled them to swivel their wrists with far more flexibility. In the short term, this gave them more dexterity as predators, but it also opened a door in the adjacent possible that would eventually lead, many millions of years later, to the evolution of wings and flight. When our ancestors evolved opposable thumbs, they opened up a whole new cultural branch of the adjacent possible: the creation and use of finely crafted tools and weapons.

One of the things that I find so inspiring in Kauffman's notion of the adjacent possible is the continuum it suggests between natural and man-made systems. He introduced the concept in part to illustrate a fascinating secular trend shared by both natural and human history: this relentless pushing back against the barricades of the adjacent possible. "Something has obviously happened in the past 4.8 billion years," he writes. "The biosphere has expanded, indeed, more or less persistently exploded, into the ever-expanding adjacent possible. . . . It is more than slightly interesting that this fact is clearly true, that it is rarely remarked upon, and that we have no particular theory for this expansion." Four billion years ago, if you were a carbon atom, there were a few hundred molecular configurations you could stumble into. Today that same carbon atom, whose atomic properties haven't changed one single nanogram, can help build a sperm whale or a giant redwood or an H1N1 virus, along with a near-infinite list of other carbon-based life forms that were not part of the adjacent possible of prebiotic earth. Add to that an equally formidable list of human concoctions that rely on carbon—every single object on the planet made of plastic, for instance—and you can see how far the kingdom of the adjacent possible has expanded since those fatty acids self-assembled into the first membrane.

The history of life and human culture, then, can be told as the story of a gradual but relentless probing of the adjacent possible, each new innovation opening up new paths to explore. But some systems are more adept than others at exploring those possibility spaces. The mystery of Darwin's paradox that we began with ultimately revolves around the question of why a coral reef ecosystem should be so adventurous in its exploration of the adjacent possible—so many different life forms sharing such a small space—while the surrounding waters of the ocean lack that same marvelous diversity. Similarly, the environments of big cities allow far more commercial exploration of the adjacent possible than towns or villages, allowing tradesmen and entrepreneurs to specialize in fields that would be unsustainable in smaller population centers.

The Web has explored the adjacent possible of its medium far faster than any other communications technology in history. In early 1994, the Web was a text-only medium, pages of words connected by hyperlinks. But within a few years, the possibility space began to expand. It became a medium that let you do financial transactions, which turned it into a shopping mall and an auction house and a casino. Shortly afterward, it became a true two-way medium where it was as easy to publish your own writing as it was to read other people's, which engendered forms that the world had never seen before: user-authored encyclopedias, the blogosphere, social network sites. YouTube made the Web one of the most influential video delivery mechanisms on the planet. And now digital maps are unleashing their own cartographic revolutions.

You can see the fingerprints of the adjacent possible in one of the most remarkable patterns in all of intellectual history, what scholars now call "the multiple": A brilliant idea occurs to a scientist or inventor somewhere in the world, and he goes public with his remarkable finding, only to discover that three other minds had independently come up with the same idea in the past year. Sunspots were simultaneously discovered in 1611 by four scientists living in four different countries. The first electrical battery was invented separately by Dean Von Kleist and Cuneus of Leyden in 1745 and 1746. Joseph Priestley and Carl Wilhelm Scheele independently isolated oxygen between 1772 and 1774. The law of the conservation of energy was formulated separately four times in the late 1840s. The evolutionary importance of genetic mutation was proposed by S. Korschinsky in 1899 and then by Hugo de Vries in 1901, while the impact of X-rays on mutation rates was independently uncovered by two scholars in 1927. The telephone, telegraph, steam engine, photograph vacuum tube, radio—just about every essential technological advance of modern life has a multiple lurking somewhere in its origin story.

In the early 1920s, two Columbia University scholars named William Ogburn and Dorothy Thomas decided to track down as many multiples as they could find, eventually publishing their survey in an influential essay with the delightful title "Are Inventions Inevitable?" Ogburn and Thomas found 148 instances of independent innovation, most them occurring within the same decade. Reading the list now, one is struck not just by the sheer number of cases, but how indistinguishable the list is from an unfiltered history of big ideas. Multiples have been invoked to support hazy theories about the "zeitgeist," but they have a much more grounded explanation. Good ideas are not conjured out of thin air; they are built out of a collection of existing parts, the composition of which expands (and, occasionally, contracts) over time. Some of those parts are conceptual: ways of solving problems, or new definitions of what constitutes a problem in the first place. Some of them are, literally, mechanical parts. To go looking for oxygen, Priestley and Scheele needed the conceptual framework that the air was itself something worth studying and that it was made up of distinct gases; neither of these ideas became widely accepted until the second half of the eighteenth century. But they also needed the advanced scales that enabled them to measure the minuscule changes in weight triggered by oxidation, technology that was itself only a few decades old in 1774. When those parts became available, the discovery of oxygen entered the realm of the adjacent possible. Isolating oxygen was, as the saying goes, "in the air," but only because a specific set of prior discoveries and inventions had made that experiment thinkable.

. . .

The adjacent possible is as much about limits as it is about openings. At every moment in the timeline of an expanding biosphere, there are doors that cannot be unlocked yet. In human culture, we like to think of breakthrough ideas as sudden accelerations on the timeline, where a genius jumps ahead fifty years and invents something that normal minds, trapped in the present moment, couldn't possibly have come up with. But the truth is that technological (and scientific) advances rarely break out of the adjacent possible; the history of cultural progress is, almost without exception, a story of one door leading to another door, exploring the palace one room at a time. But of course, human minds are not bound by the finite laws of molecule formation, and so every now and then an idea does occur to someone that teleports us forward a few rooms, skipping some exploratory steps in the adjacent possible. But those ideas almost always end up being short-term failures, precisely because they have skipped ahead. We have a phrase for those ideas: we call them "ahead of their time."

Consider the legendary Analytical Engine designed by nineteenth-century British inventor Charles Babbage, who is considered by most technology historians to be the father of modern computing, though he should probably be called the great-grandfather of modern computing, because it took several generations for the world to catch up to his idea. Babbage is actually in the pantheon for two inventions, neither of which he managed to build during his lifetime. The first was his Difference Engine, a fantastically complex fifteen-ton contraption, with over 25,000 mechanical parts, designed to calculate polynomial functions that were essential to

creating the trigonometric tables crucial to navigation. Had Babbage actually completed his project, the Difference Engine would
have been the world's most advanced mechanical calculator. When
the London Science Museum constructed one from Babbage's plans
to commemorate the centennial of his death, the machine returned
accurate results to thirty-one places in a matter of seconds. Both the
speed and precision of the device would have exceeded anything
else possible in Babbage's time by several orders of magnitude.

For all its complexity, however, the Difference Engine was well
within the adjacent possible of Victorian technology. The second half
of the nineteenth century saw a steady stream of improvements to
mechanical calculation, many of them building on Babbage's architecture. The Swiss inventor Per Georg Scheutz constructed a working Difference Engine that debuted at the Exposition Universelle of
1855; within two decades the piano-sized Scheutz design had been
reduced to the size of a sewing machine. In 1884, an American inventor named William S. Burroughs founded the American Arithmom-
eter Company to sell mass-produced calculators to businesses around
the country. (The fortune generated by those machines would help
fund his namesake grandson's writing career, not to mention his drug
habit, almost a century later.) Babbage's design for the Difference
Engine was a work of genius, no doubt, but it did not transcend the
adjacent possible of its day.

The same cannot be said of Babbage's other brilliant idea: the
Analytical Engine, the great unfulfilled project of Babbage's career,
which he toiled on for the last thirty years of his life. The machine
was so complicated that it never got past the blueprint stage, save a
small portion that Babbage built shortly before his death in 1871.
The Analytical Engine was—on paper, at least—the world's first

programmable computer. Being programmable meant that the machine was fundamentally open-ended; it wasn't designed for a specific set of tasks, the way the Difference Engine had been optimized for polynomial equations. The Analytical Engine was, like all modern computers, a shape-shifter, capable of reinventing itself based on the instructions conjured by its programmers. (The brilliant mathematician Ada Lovelace, the only daughter of Lord Byron, wrote several sets of instructions for Babbage's still-vaporware Analytical Engine, earning her the title of the world's first programmer.) Babbage's design for the engine anticipated the basic structure of all contemporary computers: "programs" were to be inputted via punch cards, which had been invented decades before to control textile looms; instructions and data were captured in a "store," the equivalent of what we now call random access memory, or RAM; and calculations were executed via a system that Babbage called "the mill," using industrial-era language to describe what we now call the central processing unit, or CPU.

Babbage had most of this system sketched out by 1837, but the first true computer to use this programmable architecture didn't appear for more than a hundred years. While the Difference Engine engendered an immediate series of refinements and practical applications, the Analytical Engine effectively disappeared from the map. Many of the pioneering insights that Babbage had hit upon in the 1830s had to be independently rediscovered by the visionaries of World War II–era computer science.

Why did the Analytical Engine prove to be such a short-term dead end, given the brilliance of Babbage's ideas? The fancy way to say it is that his ideas had escaped the bounds of the adjacent possible. But it is perhaps better put in more prosaic terms: Babbage simply didn't have the right spare parts. Even if Babbage had built a machine to his specs, it is unclear whether it would have worked, because Babbage was effectively sketching out a machine for the electronic age during the middle of the steam-powered mechanical revolution. Unlike all modern computers, Babbage's machine was to be composed entirely of mechanical gears and switches, staggering in their number and in the intricacy of their design. Information flowed through the system as a constant ballet of metal objects shifting positions in carefully choreographed movements. It was a maintenance nightmare, but more than that, it was bound to be hopelessly slow. Babbage bragged to Ada Lovelace that he believed the machine would be able to multiply two twenty-digit numbers in three minutes. Even if he was right—Babbage wouldn't have been the first tech entrepreneur to exaggerate his product's performance—that kind of processing time would have made executing more complicated programs torturously slow. The first computers of the digital age could perform the same calculation in a matter of seconds. An iPhone completes millions of such calculations in the same amount of time. Programmable computers needed vacuum tubes, or, even better, integrated circuits, where information flows as tiny pulses of electrical activity, instead of clanking, rusting, steam-powered metal gears.

You can see a comparable pattern—on a vastly accelerated timetable—in the story of YouTube. Had Hurley, Chen, and Karim tried to execute the exact same idea for YouTube ten years earlier, in 1995, it would have been a spectacular flop, because a site for sharing video was not within the adjacent possible of the early Web. For starters, the vast majority of Web users were on painfully slow dial-up connections that could sometimes take minutes to download a small image. (The average two-minute-long YouTube clip would have taken as much as an hour to download on the then-standard 14.4 bps modems.) Another key to YouTube's early success is that its developers were able to base the video serving on Adobe's Flash platform, which meant that they could focus on the ease of sharing and discussing clips, and not spend millions of dollars developing a whole new video standard from scratch. But Flash itself wasn't released until late 1996, and didn't even support video until 2002.

To use our microbiology analogy, having the idea for a Difference Engine in the 1830s was like a bunch of fatty acids trying to form a cell membrane. Babbage's calculating machine was a leap forward, to be sure, but as advanced as it was, the Difference Engine was still within the bounds of the adjacent possible, which is precisely why so many practical iterations of Babbage's design emerged in the subsequent decades. But trying to create an Analytical Engine in 1850—or YouTube in 1995—was the equivalent of those fatty acids trying to self-organize into a sea urchin. The idea was right, but the environment wasn't ready for it yet.

All of us live inside our own private versions of the adjacent possible. In our work lives, in our creative pursuits, in the organizations that employ us, in the communities we inhabit—in all these different environments, we are surrounded by potential new configurations, new ways of breaking out of our standard routines. We are, each of us, surrounded by the conceptual equivalent of those Toyota spare parts, all waiting to be recombined into something magical, something new. It need not be the epic advances of biological diversity, or the invention of programmable computing. Unlocking a new door can lead to a world-changing scientific breakthrough, but it can also lead to a more effective strategy for teaching second-graders, or a novel marketing idea for the vacuum cleaner your company's about to release. The trick is to figure out ways to explore the edges of possibility that surround you. This can be as simple as changing the physical environment you work in, or cultivating a specific kind of social network, or maintaining certain habits in the way you seek out and store information.

Recall the question we began with: What kind of environment creates good ideas? The simplest way to answer it is this: innovative environments are better at helping their inhabitants explore the adjacent possible, because they expose a wide and diverse sample of spare parts—mechanical or conceptual—and they encourage novel ways of recombining those parts. Environments that block or limit those new combinations—by punishing experimentation, by obscuring certain branches of possibility, by making the current state so satisfying that no one bothers to explore the edges—will, on average, generate and circulate fewer innovations than environments that encourage exploration. The infinite variety of life that so impressed Darwin, standing in the calm waters of the Keeling Islands, exists because the coral reef is supremely gifted at recycling and reinventing the spare parts of its ecosystem.

There's a famous moment in the story of the near-catastrophic *Apollo 13* mission—wonderfully captured in the Ron Howard film—where the mission control engineers realize they need to create an improvised carbon dioxide filter, or the astronauts will poison the lunar module atmosphere with their own exhalations before they return to Earth. The astronauts have plenty of carbon "scrubbers" on board, but these filters were designed for the original, damaged spacecraft, and don't fit the air ventilation system of the lunar module they are using as a lifeboat to return home. Mission Control quickly assembles what it calls a "tiger team" of engineers to hack their way through the problem, and creates a rapid-fire inventory of all the available equipment currently on the lunar module. In the movie, Deke Slayton, head of Flight Crew Operations, tosses a jumbled pile of gear on a conference table: suit hoses, canisters, stowage bags, duct tape, and other assorted gadgets. He holds up the carbon scrubbers. "We gotta find a way to make this fit into a hole for this," he says, and then points to the spare parts on the table, "using nothing but *that*."

The space gear on the table defines the adjacent possible for the problem of building a working carbon scrubber on a lunar module. The device they eventually concoct, dubbed the "mailbox," performs beautifully. The canisters and nozzles are like the ammonia and methane molecules of the early earth, or Babbage's mechanical gears, or those Toyota parts heating an incubator: they are the building blocks that create—and limit—the space of possibility for a specific problem. In a way, the engineers at Mission Control had it easier than most. Challenging problems don't usually define their adjacent possible in such a clear, tangible way. Part of coming up with a good idea is discovering what those spare parts are, and ensuring that you're not just recycling the same old ingredients. This, then, is where the next six patterns of innovation will take us, because they all involve, in one way or another, tactics for assembling a more eclectic collection of building block ideas, spare parts that can be reassembled into useful new configurations. The trick to having good ideas is not to sit around in glorious isolation and try to think big thoughts. The trick is to get more parts on the table.

II.

LIQUID NETWORKS

There are a dozen different metaphors we use colloquially to describe good ideas: we call them sparks, flashes, lightbulb moments; we have brainstorms and breakthroughs, eureka moments and epiphanies. Something about the concept pushes our language into rhetorical overdrive, our verbiage straining to reproduce the innovation it describes.

And yet, florid as they are, none of those metaphors captures what an idea actually is, on the most elemental level.

A good idea is a *network*. A specific constellation of neurons—thousands of them—fire in sync with each other for the first time in your brain, and an idea pops into your consciousness. A new idea is a network of cells exploring the adjacent possible of connections that they can make in your mind. This is true whether the idea in question is a new way to solve a complex physics problem, or a closing line for a novel, or a feature for a software application. If we're going to try to explain the mystery of where ideas come from, we'll have to start by shaking ourselves free of this common misconception: an idea is not a single thing. It is more like a swarm.

When you think about ideas in their native state of neural networks, two key preconditions become clear. First, the sheer size of the network: you can't have an epiphany with only three neurons firing. The network needs to be densely populated. Your brain has roughly 100 billion neurons, an impressive enough number, but all those neurons would be useless for creating ideas (as well as all the other achievements of the human brain) if they weren't capable of making such elaborate connections with each other. The average neuron connects to a thousand other neurons scattered across the brain, which means that the adult human brain contains 100 trillion distinct neuronal connections, making it the largest and most complex network on earth. (By comparison, there are somewhere on the order of 40 billion pages on the Web. If you assume an average of ten links per page, that means you and I are walking around with a high-density network in our skulls that is orders of magnitude larger than the entirety of the World Wide Web.)

The second precondition is that the network be *plastic*, capable of adopting new configurations. A dense network incapable of forming new patterns is, by definition, incapable of change, incapable of probing at the edges of the adjacent possible. When a new idea pops into your head, the sense of novelty that makes the experience so magical has a direct correlate in the cells of your brain: a brand-new assemblage of neurons has come together to make the thought possible. Those connections are built by our genes and by personal experience: some connections help regulate our heartbeat and trigger reflex reactions; others conjure up vivid sense memories of the cookies we ate as children; others help us invent the concept of a programmable computer. The connections are the key to wisdom, which is why the whole notion of losing neurons after we hit adulthood is a red herring. What matters in your mind is not just the number of neurons, but the myriad connections that have formed between them.

Of course, everything that happens in your brain is, technically speaking, a network. Remembering to cut your toenails involves a network of neurons firing in some kind of orderly fashion. But that doesn't make it an epiphany. It turns out that good ideas have certain signature patterns in the networks that make them. The creating brain behaves differently from the brain that is performing a repetitive task. The neurons communicate in different ways. The networks take on distinct shapes.

The question is how to push your brain toward those more creative networks. The answer, as it happens, is delightfully fractal: to make your mind more innovative, you have to place it inside environments that share that same network signature: networks of ideas or people that mimic the neural networks of a mind exploring the boundaries of the adjacent possible. Certain environments enhance the brain's natural capacity to make new links of association. But these patterns of connection are much older than the human brain, older than neurons even. They take us back, once again, to the origin of life itself.

As far as we know, "carbon-based life" is a redundant expression: life would be impossible without the carbon atom. Most astrobiologists—scientists who study the possibility of life elsewhere in the universe—believe that if we are ever to discover con-

vincing evidence of extraterrestrial life, be it on Mars or in some
distant galaxy, it, too, will turn out to be carbon-based.

Why are we so confident about carbon's essential role in creating living things? The answer has to do with the core properties
of the carbon atom itself. Carbon has four valence electrons resid-
ing in the outermost shell of the atom, which, for complicated reasons, makes it uniquely talented at forming connections with other
atoms, particularly with hydrogen, nitrogen, oxygen, phosphorus,
sulfur—and, crucially, with other carbon atoms. These six atoms
make up 99 percent of the dry weight of all living organisms on
earth. Those four valence bonds give carbon a strong propensity for
forming elaborate chains and rings of polymers: everything from
the genetic information stored in nucleic acids, to the building
blocks of proteins, to the energy storage of carbohydrates and fats.
(Modern technology has exploited the generative potential of the
carbon atom via the artificial polymers we call plastics.) Carbon
atoms measure only 0.03 percent of the overall composition of the
earth's crust, and yet they make up nearly 20 percent of our body
mass. That abundance highlights the unique property of the carbon
atom: its combinatorial power. Carbon is a connector.

Those connections are essential for the day-to-day functioning
of life: chains of nucleic acids instructing amino acids to assemble
into long strings of protein, powered by the stored energy of carbohydrates. But the connective properties of carbon were essential to
the original innovations of life itself. Without carbon's innate talent
for forming new complex molecules with other atoms, it is hard to
imagine how the first organisms would have evolved. Those four
valence electrons allowed the prebiotic earth to explore its own adjacent possible, sifting through the long list of potential molecular

combinations until it hit upon a series of stable chemical reactions that blossomed into the first organisms. Without the generative links of carbon, the earth would have likely remained a lifeless soup of elements, a planet of dead chemistry.

Carbon's connective talents lie at the center of one of the most famous scientific experiments of the twentieth century. In 1953, two University of Chicago professors, Stanley L. Miller and Harold C. Urey, created a closed system of glass tubes and flasks that simulated the early conditions of the prebiotic earth. The main ingredients were methane (CH_4), ammonia (NH_3), hydrogen (H_2), and water (H_2O). Only the methane contained carbon atoms. One flask connected to the chemical soup contained a pair of electrodes, which Miller and Urey used to simulate lightning by triggering a series of quick sparks between them. They ran the experiment for seven straight days, and by the time they had completed the first cycle, they found that more than 10 percent of the carbon had spontaneously recombined into many of the organic compounds essential to life: sugars, lipids, nucleic acids. Miller claimed at the time that "just turning on the spark in a basic pre-biotic experiment" produced half of the twenty-two amino acids. Several years ago, a team reanalyzed the original flasks from the Miller-Urey experiments, and found that in one version—which simulated the environment around an undersea volcano—all twenty-two amino acids had been created.

In the half-century that has passed since Miller and Urey triggered their primordial spark, hundreds of rival theories have emerged to explain the early stages of life, some emphasizing the initial development of self-replication, some emphasizing the development of metabolism; some are predicated on the intense heat of undersea vents, others on life-bearing comets colliding with the earth's surface. But all of these theories share a common motif: the combinatorial power of the carbon atom. A few researchers and science-fiction authors have speculated on an alternate scenario, where life emerges around the silicon atom. Silicon sits directly below carbon on the periodic table, and shares its four valence electrons. But silicon lacks carbon's unique versatility, its ability to form the double and triple bonds that create the long chains and rings of fatty acids and sugars. Silicon also requires far more energy to form bonds than carbon does. Tellingly, the earth contains over a hundred times as much silicon as it does carbon, and yet Mother Nature decided to base life on the much rarer element.

Silicon-based life may be impossible for one other reason: silicon bonds readily dissolve in water. Most theories of life's origin depend on H_2O not merely because hydrogen and oxygen are important elements in many organic compounds, but also because the environment of liquid water facilitated the early "chemistry experiments" that led to the emergence of life. The Miller-Urey experiment was, in a way, an attempt to test more rigorously a hunch that Charles Darwin had had a century before about the watery origins of life. In a letter to the botanist Joseph Hooker, Darwin speculated that life had first emerged in "some warm little pond, with all sorts of ammonia and phosphoric salts, light, heat, electricity." Most theories of life's origins incorporate some variation of the "primordial soup": an environment where novel combinations could occur thanks to the swirl and flow of liquid. Carbon may be a talented connector, but without a medium that allows it to collide randomly with other elements, those connective powers are likely to go to waste. All those spectacular polymer chains would remain unrealized, hidden behind the locked doors of the adjacent possible.

Like carbon, the H_2O molecule possesses several exceptional properties that make the medium of liquid water uniquely suited to sustain early life. The hydrogen bonds that form *between* distinct water molecules are about ten times stronger than equivalent bonds in "normal" liquids, which gives the medium several crucial properties. For starters, the temperature range at which water remains in liquid form is much larger than that of almost every other substance, thanks in part to those hydrogen bonds, thus preventing the oceans from boiling away during the early days of life on earth. Water is also a fiendishly talented dissolver of things. (Even the famously inert gold is soluble in seawater if you give it enough time.) The combination of water's fluidity and solubility makes it marvelously adept at creating new networks of elements, as they churn through the ever-shifting medium, colliding with each other in unpredictable ways. At the same time, the strength of the hydrogen bonds means that new combinations with some stability to them—many of them anchored around carbon atoms—can endure and seek out additional connections in the soup.

And so, when we look back to the original innovation engine on earth, we find two essential properties. First, a capacity to make new connections with as many other elements as possible. And, second, a "randomizing" environment that encourages collisions between all the elements in the system. On earth, at least, the story of life's creativity begins with a liquid, high-density network: connection-hungry carbon atoms colliding with other elements in the primordial soup. The molecules they formed mark the point at

which chemistry and physics gave way to biology. When the first
lipids self-assembled, they unlocked a door that would ultimately
lead to the cell membrane; when the first nucleotides formed, a
wing of the adjacent possible opened that eventually traced a path
to DNA. They were the first hints of life's good idea.

The computer scientist Christopher Langton observed several
decades ago that innovative systems have a tendency to gravitate toward the "edge of chaos": the fertile zone between too much
order and too much anarchy. (The notion is central to Stuart Kauffman's idea of the adjacent possible, as well.) Langton sometimes uses
the metaphor of different phases of matter—gas, liquid, solid—to
describe these network states. Think of the behavior of molecules in
each of these three conditions. In a gas, chaos rules; new configurations are possible, but they are constantly being disrupted and torn
apart by the volatile nature of the environment. In a solid, the opposite happens: the patterns have stability, but they are incapable of
change. But a liquid network creates a more promising environment
for the system to explore the adjacent possible. New configurations
can emerge through random connections formed between molecules, but the system isn't so wildly unstable that it instantly destroys
its new creations. Those connective carbon atoms swirling in the
primordial soup formed a high-density liquid network. The 100 billion neurons in your brain form another kind of liquid network:
densely interconnected, constantly exploring new patterns, but also
capable of preserving useful structures for long periods of time.

There is a prediction (albeit retroactive) lurking in this idea
of the liquid network, as well as in the premise that innovative

environments share signature patterns at different scales. The prediction is that whenever human beings first organized themselves
into settlements that resembled liquid networks, a great flowering
of innovation would have immediately followed. For ages, early
humans lived in the cultural equivalent of gaseous networks: small
packs of hunter-gatherers bouncing around the landscape, with almost no contact between groups. But the rise of agriculture changed
all that. For the first time, humans began forming groups that num-
bered in the thousands, or tens of thousands. After millennia of living in an intimate cluster of extended family, they began sharing a
space crowded with strangers. With that increase in population
came a crucial increase in the number of possible connections that
could be formed within the group. Good ideas could more readily
find their way into other brains and take root there. New forms of
collaboration became possible. Economists have a telling phrase for
the kind of sharing that happens in these densely populated environments: "information spillover." When you share a common civic
culture with thousands of other people, good ideas have a tendency
to flow from mind to mind, even when their creators try to keep
them secret. "Spillover" is the right word; it captures the essential
liquidity of information in dense settlements. As species go, *Homo
sapiens* had been on a fairly good run in the million years that led
up to the birth of agriculture: its members had invented spoken
language, art, sophisticated tools for hunting, and cooking. But until
they settled in cities, they had not figured out a way to live inside
a high-density liquid network.

What happened when they did? To grasp the magnitude of the
change, we need to put it in some kind of perspective, by measuring
the speed of innovation *before* the first cities were settled. So let us

condense seventy thousand years of innovation into a single time line, ending circa 2000 B.C., a few millennia after the first true cities formed.

Looking at the past from this perspective makes one thing clear: somewhere within a thousand years of the first cities emerging, human beings invented a whole new way of inventing. A strong correlation exists between those dense settlements and the dramatic surge in the societal innovation rate. But is there a causal relationship between the two? The chart alone cannot tell us, and we do not know enough about the specific histories of these innovations to document how essential the urban context was to their creation. But the circumstantial evidence is strong.

No doubt some ingenious hunter-gatherer stumbled across the cleansing properties of ashes mixed with animal fat, or dreamed of building aqueducts in those long eons before the rise of cities, and we simply have no record of his epiphany. But the lack of a record is exactly the point. In a low-density, chaotic network, ideas come and go. In the dense networks of the first cities, good ideas have a natural propensity to get into circulation. They spill over, and through that spilling they are preserved for future generations. For reasons we will see, high-density liquid networks make it easier for innovation to happen, but they also serve the essential function of *storing* those innovations. Before writing, before books, before Wikipedia, the liquid network of cities preserved the accumulated wisdom of human culture.

The pattern was repeated in the explosion of commercial and artistic innovation that emerged in the densely settled hill towns of Northern Italy, the birthplace of the European Renaissance. Once again, the rise of urban networks triggers a dramatic increase in the

Boats

Bows & Arrows
Mining Cloth

Sewing
Needles

Pottery

Rope

Baskets
Metalworking
Alcohol
Agriculture

Irrigation
Animal Husbandry

Cities

Alphabet
Candles
Aqueducts
Currency
Rulers
Papyrus
Sewers
Ploughs
Soap
Cement
Combs
Resevoirs
Writing
Silk
Paving
Canals
Sailing
Bread
Wheels
Furnaces

70,000 BC   60,000 BC   50,000 BC   40,000 BC   30,000 BC   20,000 BC   10,000 BC   5,000 BC   2,000 BC

flow of good ideas. It is not a coincidence that Northern Italy was the most urbanized region in all of Europe during the fourteenth and fifteenth centuries. But, in a crucial sense, the pattern of Renaissance innovation differs from that of the first cities: Michelangelo, Brunelleschi, and da Vinci were emerging from a medieval culture that suffered from too much order. If dispersed tribes of hunter-gatherers are the cultural equivalent of a chaotic, gaseous state, a culture where the information is largely passed down by monastic scribes stands at the opposite extreme. A cloister is a solid. By breaking up those information bonds and allowing ideas to circulate more freely through a wider, connected population, the great Italian innovators brought new life to the European mind.

Historians have long noted the connection between the artistic and scientific flowering of the Renaissance and the formation of early merchant capitalism in the region, which of course involved its own set of innovations in banking, accounting, and insurance. To be sure, capitalism accelerated the growth of the Italian cities, and created surplus wealth that was then deployed to support artists and architects like Michelangelo and Brunelleschi. But the connection between capitalism and innovation is more subtle than we often make it out to be. Yes, free markets introduce new forms of competition and capital accumulation that can drive the creation and adoption of new ideas. But markets should not be exclusively defined in terms of the profit motive. Consider the invention of one of capitalism's key conceptual tools: double-entry accounting, which Goethe called one of the "finest inventions of the human mind." Now the cornerstone of all financial bookkeeping, double-entry's innovation of recording every financial event in two ledgers (one reflecting a debit, the other a credit) allowed merchants to track the financial health of their businesses with unparalleled accuracy. First codified by the Franciscan friar and mathematician Luca Pacioli in 1494, the double-entry method had been used for at least two centuries by Italian bankers and merchants. We do not know if the method originated in the mind of a single visionary proto-accountant, or whether the idea emerged simultaneously in the minds of multiple entrepreneurs, or whether it was passed on by Islamic entrepreneurs who may have experimented with the technique centuries before. Whatever its roots, the technique first became commonplace in the trade capitals of Italy—Genoa, Venice, and Florence—as the merchants of the early Renaissance shared tips among themselves on how best to manage their finances. What makes the history of double-entry so fascinating is the simple fact that no one seems to have claimed ownership of the technique, despite its immense value to a capitalist enterprise. One of the essential instruments in the creation of modern capitalism appears to have been developed collectively, circulating through the liquid networks of Italy's cities. Double-entry accounting made it far easier to keep track of what you owned, but no one owned double-entry accounting itself. The idea was too powerful not to spill over into other nearby minds.

Double-entry accounting illustrates a key principle in the emergence of markets: when economic systems shift from feudal structures to the nascent forms of modern capitalism, they become less hierarchical and more networked. A society organized around marketplaces, instead of castles or cloisters, distributes decision-making authority across a much larger network of individual minds. The innovation power of the marketplace derives, in part, from this most elemental math: no matter how smart the "authorities" may be, if they are outnumbered a thousand to one by the marketplace, there will be more good ideas lurking in the market than in the feudal castle. Cities and markets recruit more minds into the collective project of exploring the adjacent possible. As long as there is spillover between those minds, useful innovations will be more likely to appear and spread through the population at large.

In thinking about networked innovation this way, I am specifically not talking about a "global brain," or a "hive mind." There are indeed some problems that are wonderfully solved by collective thinking: the formation of neighborhoods in cities, the variable signals of market pricing, the elaborate engineering feats of the social insects. But as many critics have pointed out—most recently, the computer scientist and musician Jaron Lanier—large collectives are rarely capable of true creativity or innovation. (We have the term "herd mentality" for a reason.) When the first market towns emerged in Italy, they didn't magically create some higher-level group consciousness. They simply widened the pool of minds that could come up with and share good ideas. This is not the wisdom of the crowd, but the wisdom of *someone* in the crowd. It's not that the network itself is smart; it's that the individuals get smarter because they're connected to the network.

In 1964, Arthur Koestler published his epic account of innovation's roots, *The Act of Creation*. The book was an attempt to explain how breakthrough ideas in science and art come about. (Koestler also had a long opening section on humor, which he believed was closely related to the more erudite inspiration of the poets and physicists.) Koestler's survey extends from Archimedes to

Einstein, from Milton to Joyce, and his analysis is unfailingly interesting, often brilliant. Yet across such a wide-ranging survey, one
pattern recurs with a surprising regularity. The act of creation, in
Koestler's account, is something that happens exclusively in the
mind. He spends almost no time discussing the many habitats that
sustain or encourage innovation. The book's index, for instance,
lacks a single reference to that great engine of supercreativity, the
city. Koestler was a great believer in the creative power that emerges
when different intellectual disciplines collide. But he seems to have
had little interest in the environments that make those collisions
possible: living environments, office environments, media environments. On a basic level, it is true that ideas happen *inside* minds,
but those minds are invariably connected to external networks that
shape the flow of information and inspiration out of which great
ideas are fashioned.

Koestler was hardly alone in his interest in the roots of scientific
breakthrough. Thomas Kuhn's even more influential book *The Structure of Scientific Revolutions* had been published two years before
The Act of Creation. Since those two books were published, countless dissertations and scholarly essays have explored the psychology
and sociology of scientific progress. Some focused on biographical
accounts of legendary scientists at work; others tested theories
through lab experiments that simulated the kind of cognitive work
involved in scientific discovery. Others conducted extensive interviews with prominent researchers, asking them to recall the details
of their eureka moments and private paradigm shifts.

In the early 1990s, a psychologist at McGill University named
Kevin Dunbar decided to take another approach: instead of reading
biographies or theorizing in the lab or listening to scientists recount

their greatest hits, he would actually watch them as they worked. Dunbar's research style was closer to the reality show *Big Brother* than it was to traditional philosophy of science. He set up cameras in four leading molecular biology laboratories and recorded as much of the action as possible. He also conducted extensive interviews in which the researchers described the latest developments in their experiments and their shifting hypotheses, all in the present tense. The taping and *in medias res* interviews allowed Dunbar to get around one of the major failings of traditional studies that rely on retrospective interviews: people tend to condense the origin stories of their best ideas into tidy narratives, forgetting the messy, convoluted routes to inspiration that they actually followed. Dunbar called his approach *in vivo*, as opposed to the more traditional *in vitro* approach to studying scientific cognition. In other words, Dunbar wasn't studying idea formation in an artificial petri dish environment. He was studying it in the wild.

Dunbar and his team transcribed all the interactions and coded each exchange using a classification scheme that allowed them to track patterns in the flow of information through the lab. In group interactions, for instance, exchanges between scientists could be formally coded as "clarification" or "agreement and elaboration" or "questioning." Most important, Dunbar tracked the conceptual changes that occurred over the course of each project: a researcher baffled by persistent problems in achieving a stable control result who suddenly realizes that the control problem could be the basis for a whole new experiment; an exchange between two scientists working on different projects who recognize a surprising and important connection between their work.

The most striking discovery in Dunbar's study turned out to

be the physical location where most of the important breakthroughs occurred. With a science like molecular biology, we inevitably have an image in our heads of the scientist alone in the lab, hunched over a microscope, and stumbling across a major new finding. But Dunbar's study showed that those isolated eureka moments were rarities. Instead, most important ideas emerged during regular lab meetings, where a dozen or so researchers would gather and informally present and discuss their latest work. If you looked at the map of idea formation that Dunbar created, the ground zero of innovation was not the microscope. It was the conference table.

Dunbar uncovered a set of interactions that consistently led to important breakthroughs during lab conversations. The group environment helped recontextualize problems, as questions from colleagues forced researchers to think about their experiments on a different scale or level. Group interactions challenged researchers' assumptions about their more surprising findings, making them less likely to dismiss them as experimental error. In group problem-solving sessions, Dunbar writes, "the results of one person's reasoning became the input to another person's reasoning . . . resulting in significant changes in all aspects of the way the research was conducted." Productive analogies between different specialized fields were more likely to emerge in the conversational setting of the lab meeting.

Dunbar's research suggests one vaguely reassuring thought: even with all the advanced technology of a leading molecular biology lab, the most productive tool for generating good ideas remains a circle of humans at a table, talking shop. The lab meeting creates an environment where new combinations can occur, where information can spill over from one project to another. When you work alone in an office, peering into a microscope, your ideas can get trapped in place, stuck in your own initial biases. The social flow of the group conversation turns that private solid state into a liquid network.

D unbar's generative conference room meetings remind us that the physical architecture of our work environments can have a transformative effect on the quality of our ideas. The quickest way to freeze a liquid network is to stuff people into private offices behind closed doors, which is one reason so many Web-era companies have designed their work environments around common spaces where casual mingling and interdepartmental chatter happens without any formal planning. (In a *New Yorker* essay, Malcolm Gladwell wonderfully described this trend as the West Village–ification of the corporate office.) The idea, of course, is to strike the right balance between order and chaos. Inspired by the early hype about telecommuting, the advertising agency TBWA/Chiat/Day experimented with a "nonterritorial" office where desks and cubicles were jettisoned, along with the private offices: employees had no fixed location in the office and were encouraged to cluster in new, ad hoc configurations with their colleagues depending on that day's projects. By all accounts, it was a colossal failure, precisely because it traded excessive order for excessive chaos.

Slightly less ambitious open-office plans have grown increasingly unfashionable in recent years, for one compelling reason: people don't like to work in them. To work in an open office is to work exclusively in public, which turns out to have just as many drawbacks as working entirely in your private lab. A better model might be MIT's legendary Building 20, the temporary structure built during World War II that somehow managed to last fifty-five years, in part because it had an extraordinary track record for cultivating both breakthrough ideas and organizations like Noam Chomsky's linguistics department, Bose Acoustics, and the Digital Equipment Corporation. As MIT wrote in a press release commemorating the building's remarkable history: "Not assigned to any one school, department, or center, it seems to always have had space for the beginning project, the graduate student's experiment, the interdisciplinary research center."

The magic of Building 20, powerfully eulogized in Stewart Brand's *How Buildings Learn*, lay in the balance the environment struck between order and chaos. There were walls and doors and offices, as in most academic buildings. But the structure's temporary origins—it was originally built with the expectation that it would be torn down after five years—meant that those structures could be reconfigured with little bureaucratic fuss, as new ideas created new purposes for the space.

Because they are fixed physical structures, most offices have a natural tendency to disrupt liquid networks of information. They themselves are, quite literally, made out of solids, and they often map out the conceptual solid of a formal org chart, with its neatly defined departments and hierarchies. Building 20 resisted those calcifying forces for a simple reason: it was built on the cheap, which meant its residents had no qualms about tearing down a wall or punching a hole in the ceiling to adapt the space to a new idea. But architects and interior designers are learning how to build work environments that facilitate liquid networks in more permanent structures.

In November of 2007, Microsoft opened the doors to the new

Redmond, Washington–based headquarters of its research division: Building 99. Created by a Microsoft designer named Martha Clarkson after deep collaboration with the tinkerers and multidisciplinarians of the research division, Building 99 was created from the ground up to be reinvented by the unpredictable flow of collaboration and inspiration. All the office spaces are modular, with walls that can be easily reconfigured to match the needs of the employees. Larger "situation rooms" house groups working on high-priority projects, with a mix of private workstations, conference tables, and sofas. Most walls are write-on/wipe-off, so if inspiration hits on the way to the restroom, you can quickly sketch out an idea for your colleagues to see. The traditional kitchenette with a coffeepot and refrigerator is replaced by open "mixer stations" where employees gather to share ideas or gossip. In a sense, Clarkson built the watercoolers first, and then designed an office building around them.

Two decades ago, the psychologist Mihaly Csikszentmihalyi proposed the concept of "flow" to describe the internal state of energized focus that characterizes the mind at its most productive. It's a lovely metaphor, precisely because it suggests the essential fluidity that good ideas so often need. Flow is not the singular intensity of focusing "like a laser," as we often say. And it is not the miraculous illumination of a sudden brainstorm. Rather, it is more the feeling of drifting along a stream, being carried in a clear direction, but still tossed in surprising ways by the eddies and whirls of moving water.

But standing in the atrium of Building 99, it's impossible not to think that this space was designed to conjure up a different kind of flow: the collective flow of energized minds forming liquid networks in their mixing spaces and situation rooms. Building 99—like

Building 20 before it—is a space that sees information spillover as a feature, not a flaw. It is designed to leak. In this, it shares some core values with the liquid networks of dense cities. No, a closed office at one of the world's richest corporations will never have the open-ended collisions and vitality that a city sidewalk has. But those are extreme points on a continuum. What is important in a structure like Building 99 is what it has learned about flow from those urban environments, and from temporary structures like Building 20. A corporate office building will never re-create fourteenth-century Genoa, or even twentieth-century Greenwich Village. But office design is moving in that direction, away from the crystal palaces of Organization Man, with their corner offices and anonymous cubicles. And with that increased fluidity—all those new ideas jostling against each other, in rooms expanding and contracting to meet their needs—it's not hard to imagine the space generating a reliable flow of innovation in the years to come. Exploring the adjacent possible can be as simple as opening a door. But sometimes you need to move a wall.

III.

THE SLOW HUNCH

On July 10, 2001, an Arizona-based FBI field agent named Ken Williams filed an "electronic communication" with his superiors in Washington and New York, using the Bureau's Automated Case Support system, the antiquated electronic repository through which the Bureau shared information about ongoing investigations. The six-page document began with this prophetic sentence: "The purpose of this communication is to advise the Bureau and New York of the possibility of a coordinated effort by USAMA BIN LADEN (UBL) to send students to the United States to attend civil aviation universities and colleges."

This was the now legendary "Phoenix memo," a warning shot fired—and largely ignored—during the lazy summer months leading up to 9/11. (Ironically, the very day that Williams filed his memo, the *New York Times* ran an op-ed titled "The Declining Terrorist Threat.") Williams had been inspired to write the memo by a pattern he had detected over the preceding year: an "inordinate number" of people of "investigative interest" who had registered for various flight schools and other civil aviation colleges in Arizona. Williams had conducted interviews with several of these subjects, including one Zakaria Mustapha Soubra, an aeronautical engineering student on an F-1 visa from the UK. Soubra had pictures of bin Laden in his home and told Williams that he believed the U.S. forces and embassies attacked in the Gulf and in Africa had been "legitimate military targets of Islam." Williams also suggested that nine other students from Algeria, Kenya, India, Saudi Arabia, and other Middle Eastern states had enrolled in flight schools and possessed extensive ties to radical Islamic movements. Two of them apparently were acquaintances of Hani Hanjour, who would be at the controls of American Airlines Flight 77 when it crashed into the Pentagon on the morning of September 11.

Williams failed to anticipate the immediacy of the threat; the memo suggests a long-term plan that would "establish a cadre of individuals who will one day be working in the civil aviation community around the world. These individuals will be in a position in the future to conduct terror activity against civil aviation targets." Williams thought that al Qaeda might be plotting a measured infiltration of the airline industry; he failed to imagine the brute-force hijacking that was to unfold just two months later. But his recommendations were right on target. Williams argued that the FBI should assemble a comprehensive list of all flight schools and other aviation institutions around the country, and flag anyone attempting to obtain a visa to attend one of these schools.

Though it had been addressed to several high-level offices,

including that of David Frasca, the head of the Radical Fundamentalist Unit in D.C., Williams's memo quickly entered what investigators later dubbed a "black hole" at FBI headquarters. For nearly
three weeks, it remained in limbo, before it was finally assigned to
an analyst to review. The analyst labeled it "routine" instead of
"urgent." Another agent in New York called it "speculative and not
very significant." Though it was standard for analysts to then pass
on reports of this ilk to their superiors, the memo never reached
RFU chief Frasca.

When word of the memo first leaked during 2002, intelligence
and law enforcement officials were quick to dismiss Williams's
warning, calling it nothing more than a hunch. "He made a recom-
mendation that we initiate a program to look at flight schools that
was received at Headquarters," FBI chief Robert Mueller testified.
"It was not acted on by September 11. I should say in passing that
even if we had followed those suggestions at that time, it would not,
given what we know since September 11, have enabled us to prevent
the attacks of September 11."

Both statements about the Phoenix memo are demonstrably
true. Williams had a hunch about terrorist groups and flight schools,
and that hunch on its own would not have been enough to prevent
the attacks of September 11. But dismissing it on those grounds
fundamentally misses the point. Williams stumbled across a provocative and surprising idea that was nonetheless incomplete. But
if that hunch had connected with another equally provocative idea,
one that emerged three weeks later and five hundred miles away,
the Phoenix memo might well have transformed the history of the
early twenty-first century.

. . .

You can learn a great deal about the history of innovation by examining great ideas that changed the world. Indeed, most intellectual histories are structured in exactly this fashion, a narrative of breakthroughs and insights and eureka moments that had a transformative impact on human society. But because those ideas were by definition successful ones, it's tempting to attribute their success to intrinsic causes: the sheer brilliance of the idea itself, or the sheer brilliance of the mind that came up with it. But those intrinsic causes can easily overshadow the environmental role in the creation and spread of those ideas. This is why it is just as useful to look at the sparks that failed, the ideas that found their way to a promising region of the adjacent possible but somehow collapsed there. The Phoenix memo was precisely one of those failed sparks. It contained great wisdom and foresight—in July of 2001, Ken Williams was probably closer to the 9/11 plot than any human being on the planet, save the perpetrators themselves—but that information proved to be ultimately useless. Why?

The simple answer is that no one implemented Williams's recommendations, in part because the memo itself had failed to persuade the mid-tier analysts of its importance, and in part because a communications failure inside the FBI kept the memo from reaching the top brass at Counterterrorism and the RFU. But even if the memo had reached David Frasca in mid-July, and somehow persuaded him that Ken Williams was on to something, it almost certainly would have failed to stop the 9/11 plot, because it would have taken months to cross-reference all the visa applications with the enrollment records for flight schools across the country. Detecting such subtle patterns in real time was the unrealized goal of the much-criticized Total Information Awareness project spearheaded by Admiral John Poindexter in the years immediately after 9/11. But in 2001, FBI agents could barely send e-mail to each other, much less cross-reference visa applications with flight school attendance records. This is the technicality that allowed Robert Mueller to testify that following the recommendations of the Phoenix memo would have done nothing to stop the 9/11 attacks. Looking for unusual visa applications in flight school attendees might well have led the Bureau to the hijackers, but there was no information architecture in place that could have successfully executed that kind of query in a matter of weeks. And so, by that standard, Ken Williams's hunch was not enough on its own to prevent 9/11.

But the Phoenix memo might well have been instrumental in stopping the attacks had it followed a pattern that recurs throughout the history of world-changing ideas. It was a hunch that needed to collide with another hunch.

Exactly one month after Ken Williams filed his memo, Zacarias Moussaoui enrolled at Pan Am International Flight Academy on the outskirts of St. Paul, Minnesota, where he began training to fly a Boeing 747-400 on a simulator. Instructors and other employees at the flight school were immediately suspicious about their new pupil, who paid his entire $8,300 fee in cash. Moussaoui possessed an inordinate amount of interest in the operation of the cockpit doors and flight tour communications, despite the fact that he claimed no interest in ever flying a real plane. The Pan Am employees contacted the FBI, and after a quick background check,

Moussaoui was arrested on immigration violations at a motel on August 16. An interrogation convinced the field agents, led by Harry Samit and Greg Jones, that Moussaoui posed an active threat and might be part of a wider conspiracy. The Minnesota office of the FBI then began a frantic, and ultimately unsuccessful, attempt to obtain a search warrant to examine the files on Moussaoui's laptop. On August 21, the request to seek a search warrant was formally denied on the grounds that the evidence for probable cause was "shaky," just another hunch from the hinterlands. For the next week, the Minneapolis office implored headquarters to get access to Moussaoui's laptop, to no avail. Agent Jones at one point warned that Moussaoui might "try to fly something into the World Trade Center." The search warrant would not be granted until the afternoon of September 11, after Jones's vision turned out to be all too prescient.

This is a story of two hunches: Ken Williams's hunch that a plot involving multiple radical Islamic fundamentalists could be intercepted by tracking visa applications and flight school enrollment records; and the Minneapolis field agents' hunch that Moussaoui wanted to fly a plane into the World Trade Center. (The latter began, of course, with another hunch: the Pan Am school instructors' hunch that Zacarias Moussaoui was not being honest about his interest in using a 747 simulator.) On their own, they were indeed hunches; on their own, the evidence for their validity was indeed shaky. But contemplating them together amplified their persuasive power dramatically. Connecting the dots between them would have certainly supplied enough probable cause to justify examining the contents of Zacarias Moussaoui's laptop. And had the agents examined his belongings, they would have uncovered direct connections to eleven of the 9/11 hijackers, along with Western Union wire-transfer numbers tracking recent payments from Ramzi bin al-Shibh, one of the central coordinators of the 9/11 attack. We cannot know for certain whether that information alone would have led the authorities to Mohamed Atta in time, or whether a more aggressive interrogation of Moussaoui himself might have elicited a confession that would have unraveled the plot. Certainly it is within the realm of possibility. What is undeniable is that in late August of 2001, the only real hope for stopping the attacks lay in connecting these two hunches.

The failed spark of the Phoenix memo suggests an answer to the mystery of superlinear scaling in cities and on the Web. A metropolis shares one key characteristic with the Web: both environments are dense, liquid networks where information easily flows along multiple unpredictable paths. Those interconnections nurture great ideas, because most great ideas come into the world half-baked, more hunch than revelation. Genuine insights are hard to come by; it's challenging to imagine a terrorist plot to fly passenger planes into buildings, or to invent a programmable computer. And so, most great ideas first take shape in a partial, incomplete form. They have the seeds of something profound, but they lack a key element that can turn the hunch into something truly powerful. And more often than not, that missing element is somewhere else, living as another hunch in another person's head. Liquid networks create an environment where those partial ideas can connect; they provide a kind of dating service for promising hunches. They make it easier to disseminate good ideas, of course, but they also do something more sublime: they help *complete* ideas.

The real problem with Ken Williams's hunch was not that it

failed to envision the exact details or the imminence of the 9/11 plot, or even that his recommendations would have failed to prevent the plot had they been followed. The problem with Ken Williams's hunch was *environmental*: instead of circulating through a dense network, the Phoenix memo was dropped into the black hole of the Automated Case Support system. Instead of seeking out new connections, the Phoenix memo was deposited in the equivalent of a locked file cabinet. Hunches that don't connect are doomed to stay hunches.

There is a fundamental difference between the Phoenix and Minnesota hunches, though, and that difference is *time*. The flight instructors had a bad feeling about Moussaoui in a matter of hours; something just seemed unsettling about his manner and the questions he asked. Ken Williams, on the other hand, developed his hunch about the flight school threat over years of investigation. The Phoenix memo was not the result of a gut impression; it was an idea that slowly took shape over time, a pattern detected after countless hours of observation and inquiry.

The Minnesota hunch has become intellectually fashionable in recent years: the gut instinct, the "emotional brain" flash assessment of a situation that defies the slower calculations of logic—but which nonetheless turns out to be uncannily accurate. The interest in this kind of hunch dates back to the 1980s and António Damásio's experiments with brain-damaged patients whose inability to make intuitive snap judgments produced startlingly irrational behavior. Malcolm Gladwell's bestseller *Blink* focused almost exclusively on the power (and the occasional danger) of the instant hunch: the art historian who knows in a second that the ancient sculpture is a fraud; the NYPD officer who makes a disastrous snap judgment that a suspect is reaching for a gun when he is actually reaching for his wallet.

But the snap judgments of intuition—as powerful as they can be—are rarities in the history of world-changing ideas. Most hunches that turn into important innovations unfold over much longer time frames. They start with a vague, hard-to-describe sense that there's an interesting solution to a problem that hasn't yet been proposed, and they linger in the shadows of the mind, sometimes for decades, assembling new connections and gaining strength. And then one day they are transformed into something more substantial: sometimes jolted out by some newly discovered trove of information, or by another hunch lingering in another mind, or by an internal association that finally completes the thought. Because these slow hunches need so much time to develop, they are fragile creatures, easily lost to the more pressing needs of day-to-day issues. But that long incubation period is also their strength, because true insights require you to think something that no one has thought before in quite the same way. Flash judgments are often just that— judgments. Is this guy trustworthy or not? Is the sculpture a fake? A new idea is something larger than that: it's a new perspective on a problem, or a recognition of a new opportunity that has gone unexplored to date. Those kinds of breakthroughs usually take time to develop. When the eighteenth-century scientist Joseph Priestley decided to isolate a mint sprig in a sealed glass in an ingenious experiment that ultimately proved that plants were creating oxygen— one of the founding discoveries of modern ecosystem science—he was building on a hunch that he'd been cultivating for twenty years, dating back to his boyhood obsession with trapping spiders in glass jars. He'd had a hunch that there was something interesting in the way that organisms perished when you sealed them in closed vessels, something that pointed to a larger truth. And he kept that hunch alive until he was ready to make sense of it. This was not a matter of doggedly pursuing a single line of inquiry. During those twenty years, Priestley dabbled in a dozen different fields, concocted hundreds of novel experiments in his home lab, engaged in extensive conversations with the leading intellectuals of the day. A minuscule percentage of that time was devoted directly to the problem of plant respiration. He just kept it alive in the back of his mind. Sustaining the slow hunch is less a matter of perspiration than of *cultivation.* You give the hunch enough nourishment to keep it growing, and plant it in fertile soil, where its roots can make new connections. And then you give it time to bloom.

The Vaseline-daubed lens of hindsight tends to blur slow hunches into eureka moments. Inventors, scientists, entrepreneurs, artists—they all like to tell the stories of their great breakthroughs as epiphanies, in part because there is a kind of narrative thrill that comes from that lightbulb moment of sudden clarity, and in part because the leisurely background evolution of the slow hunch is much harder to convey. But if one examines the intellectual fossil record closely, the slow hunch is the rule, not the exception.

In a famous passage from his *Autobiography*, Darwin describes his great moment of insight as a young man struggling to understand the evolution of life:

In October 1838, that is, fifteen months after I had begun my systematic enquiry, I happened to read for amusement Malthus on Population, and being well prepared to appreciate the struggle for existence which everywhere goes on from long-continued observation of the habits of animals and plants, it at once struck me that under these circumstances favourable variations would tend to be preserved, and unfavourable ones to be destroyed. The result of this would be the formation of new species. Here, then, I had at last got a theory by which to work.

This is evolution's version of Newton's apple: Malthus falls out of a tree and hits Darwin on the head, and *voilà*—natural selection is born. In part, the appeal of this eureka story stems from the simple elegance of the theory itself. Unlike more technically intricate scientific breakthroughs, it seems somehow appropriate that the basic evolutionary algorithm should just pop into the mind in a moment of recognition. (Darwin's great supporter, T. H. Huxley, is said to have exclaimed, on hearing the natural selection argument for the first time, "How incredibly stupid not to think of that.") Darwin's account also possesses a strangely poetic symmetry, because years later, when Alfred Russel Wallace independently hit upon the theory of natural selection, he claimed his breakthrough had been inspired by Malthus as well.

For almost a century, the Malthusian epiphany was the canonical story of Darwinism's roots. But in the early 1970s, a psychologist and intellectual historian named Howard Gruber decided to revisit Darwin's copious notebooks from the period, reconstructing the elaborate dance of speculation, fact-marshaling, and internal debate that led to Darwin's breakthrough in the fall of 1838. What Gruber found in the notebooks was a story very different from the account relayed in Darwin's *Autobiography*. All the core elements of Darwin's theory are present in the notebooks well before the Malthusian epiphany, which the notebooks explicitly date at September 28, 1838. Darwin understands the importance of variation; the connection between natural and artificial selection; the competition among different species for survival; the clear physiological connections among species; the epic scale of evolutionary time. All these key concepts are discussed at great length in the notebooks from 1837 on. It is not merely that Darwin possesses the puzzle pieces but fails to put them together in the right configuration. In a number of remarkable passages, written many months before the Malthusian insight, he appears to be describing the theory of natural selection in almost full dress. Exactly a year before his Malthus reading, he asks, in shorthand English: "Whether every animal produces in course of ages ten thousand varieties (influenced itself perhaps by circumstances) and those alone preserved which are well adapted?" All it takes to cement a working theory of natural selection is to modify the formula ever so slightly, and clarify that the preservation of "well adapted" forms comes from their reproductive success. And yet somehow Darwin fails to understand that he has the solution at his fingertips, and continues his enquiry for another year before "getting a theory by which to work."

Even after the Malthusian insight, Darwin seems incapable of grasping the full consequences of the theory he has established. The journal entries on September 28 are suitably excited, and do seem to grapple with the fundamental elements of the theory:

Population is increase at geometrical ratio in FAR SHORTER time than 25 . . . The final cause of all this wedging, must be to sort out proper structure, and adapt it to change.——to do that for form, which Malthus shows is the final effect (by means however of volition) of this populousness on the energy of man. One may say there is a force like a hundred thousand wedges trying [to] force every kind of adapted structure into the gaps in the economy of nature, or rather forming gaps by thrusting out weaker ones.

But in the days and weeks that follow, Darwin's notes do not suggest a mind that has crossed an intellectual watershed. As Gruber notes, the very next day Darwin writes a long entry on the sexual curiosity of primates that appears to have nothing to do with his new discovery. More than a month passes before he even attempts to write down the governing rules of natural selection.

All of which means we cannot say definitively that Darwin hit upon the idea for his theory of natural selection on September 28, 1838. The best we can do is say that he did not possess the idea when he embarked on his enquiry in the summer of 1837, and that he had it in an enduring form by November of 1838. This is not a matter of gaps in the historical record. It is simply hard to pinpoint exactly when Darwin had the idea, because the idea didn't arrive in a flash; it drifted into his consciousness over time, in waves. In the months before the Malthus reading, we could probably say that Darwin had the idea of natural selection in his head, but at the same time was incapable of fully thinking it. This is how slow hunches often mature: by stealth, in small steps. They fade into view.

This pattern recurs with the other iconographic story of Darwin's intellectual journey: his formative months observing the strange diversity of the Galápagos Islands during the voyage of the *Beagle*. To be sure, Darwin's early exploration of the principles of natural selection relied heavily on the striking deviations he had seen between related species on the Galápagos archipelago. Darwin's finches are famous for a reason. But the notebooks written during the Galápagos expedition in October of 1835 have almost no hint of the world-changing theory that they will eventually inspire. In fact, the overwhelming majority of notes from Darwin's sojourn on the Galápagos are geological in nature, far more concerned with Lyell's uniformitarian theory than with the birds and reptiles of the archipelago. (One inventory of Darwin's notebooks found 1,383 pages of geological notes, versus 368 pages on zoology.) He does take extensive notes in his "naturalist" mode, but all the speculative energy of the *Beagle* journals is generated by the geology. For Darwin the biologist, the Galápagos days were a fact-finding mission, but Darwin the geologist was consciously processing and interpreting the facts as he gathered them.

According to Darwin's own account, he didn't really lock on to the tantalizing puzzle of the finches and their exotic neighbors until the next spring, just as the *Beagle* was finding safe harbor in the Keeling Islands. His journal for 1837 includes the line: "In July opened first notebook on 'Transmutation of Species'—Had been greatly struck from about Month of previous March on character of S. American fossils—& species on Galápagos archipelago. These facts origin (especially latter) of all my views." He had witnessed firsthand the marvelous diversity of species on the Galápagos, and had documented it with a precision that no human had ever attempted before. But it took him five months to realize why it was important.

. . .

Keeping a slow hunch alive poses challenges on multiple scales. For starters, you have to preserve the hunch in your own memory, in the dense network of your neurons. Most slow hunches never last long enough to turn into something useful, because they pass in and out of our memory too quickly, precisely because they possess a certain murkiness. You get a feeling that there's an interesting avenue to explore, a problem that might someday lead you to a solution, but then you get distracted by more pressing matters and the hunch disappears. So part of the secret of hunch cultivation is simple: write everything down.

We can track the evolution of Darwin's ideas with such precision because he adhered to a rigorous practice of maintaining notebooks where he quoted other sources, improvised new ideas, interrogated and dismissed false leads, drew diagrams, and generally let his mind roam on the page. We can see Darwin's ideas evolve because on some basic level the notebook platform creates a cultivating space for his hunches; it is not that the notebook is a mere transcription of the ideas, which are happening offstage somewhere in Darwin's mind. Darwin was constantly rereading his notes, discovering new implications. His ideas emerge as a kind of duet between the present-tense thinking brain and all those past observations recorded on paper. Somewhere in the middle of the Indian Ocean, a train of association compels him to revisit his notes on the fauna of the Galápagos archipelago from five months before. As he reads through his observations, a new thought begins to take shape in his mind, which provokes a whole new set of notes that will only make complete sense to Darwin two years later, after the Malthus episode.

Darwin's notebooks lie at the tail end of a long and fruitful tradition that peaked in Enlightenment-era Europe, particularly in England: the practice of maintaining a "commonplace" book. Scholars, amateur scientists, aspiring men of letters—just about anyone with intellectual ambition in the seventeenth and eighteenth centuries was likely to keep a commonplace book. The great minds of the period—Milton, Bacon, Locke—were zealous believers in the memory-enhancing powers of the commonplace book. In its most customary form, "commonplacing," as it was called, involved transcribing interesting or inspirational passages from one's reading, assembling a personalized encyclopedia of quotations. There is a distinct self-help quality to the early descriptions of commonplacing's virtues: maintaining the books enabled one to "lay up a fund of knowledge, from which we may at all times select what is useful in the several pursuits of life."

John Locke first began maintaining a commonplace book in 1652, during his first year at Oxford. Over the next decade he developed and refined an elaborate system for indexing the book's content. Locke thought his method important enough that he appended it to a printing of his canonical work, *An Essay Concerning Human Understanding*. Locke's approach seems almost comical in its intricacy, but it was a response to a specific set of design constraints: creating a functional index in only two pages that could be expanded as the commonplace book accumulated more quotes and observations:

When I meet with any thing, that I think fit to put into my common-place-book, I first find a proper head. Suppose for example that the head be EPISTOLA, I look unto the index for the first letter and the following vowel which in this instance are E. i. if in the space marked E. i. there is any number that directs me to the page designed for words that begin with an E and whose first vowel after the initial letter is I, I must then write under the word Epistola in that page what I have to remark.

Locke's method proved so popular that a century later, an enterprising publisher named John Bell printed a notebook entitled "Bell's Common-Place Book, Formed generally upon the Principles Recommended and Practised by Mr Locke." The book included eight pages of instructions on Locke's indexing method, a system which not only made it easier to find passages, but also served the higher purpose of "facilitat[ing] reflexive thought." Bell's volume would be the basis for one of the most famous commonplace books of the late eighteenth century, maintained from 1776 to 1787 by Erasmus Darwin, Charles's grandfather. At the very end of his life, while working on a biography of his grandfather, Charles obtained what he called "the great book" from his cousin Reginald. In the biography, the younger Darwin captures the book's marvelous diversity: "There are schemes and sketches for an improved lamp, like our present moderators; candlesticks with telescope stands so as to be raised at pleasure to any required height; a manifold writer; a knitting loom for stockings; a weighing machine; a surveying machine; a flying bird, with an ingenious escapement for the movement of the wings, and he suggests gunpowder or compressed air as the motive power."

The tradition of the commonplace book contains a central tension between order and chaos, between the desire for methodical arrangement, and the desire for surprising new links of association. For some Enlightenment-era advocates, the systematic indexing of the commonplace book became an aspirational metaphor for one's own mental life. The dissenting preacher John Mason wrote in 1745:

> Think it not enough to furnish this Store-house of the Mind with good Thoughts, but lay them up there in Order, digested or ranged under proper Subjects or Classes. That whatever Subject you have Occasion to think or talk upon you may have recourse immediately to a good Thought, which you heretofore laid up there under that Subject. So that the very Mention of the Subject may bring the Thought to hand; by which means you will carry a regular Common Place-Book in your Memory.

Others, including Priestley and both Darwins, used their commonplace books as a repository for a vast miscellany of hunches. The historian Robert Darnton describes this tangled mix of writing and reading:

> Unlike modern readers, who follow the flow of a narrative from beginning to end, early modern Englishmen read in fits and starts and jumped from book to book. They broke texts into fragments and assembled them into new patterns by transcribing them in different sections of their notebooks. Then they reread the copies and rearranged the patterns while adding more excerpts. Reading and writing were therefore inseparable activities. They belonged to a continuous effort to make sense of things, for the world was full of signs: you could read your

way through it; and by keeping an account of your readings, you made a book of your own, one stamped with your personality.

Each rereading of the commonplace book becomes a new kind of revelation. You see the evolutionary paths of all your past hunches: the ones that turned out to be red herrings; the ones that turned out to be too obvious to write; even the ones that turned into entire books. But each encounter holds the promise that some long-forgotten hunch will connect in a new way with some emerging obsession. The beauty of Locke's scheme was that it provided just enough order to find snippets when you were looking for them, but at the same time it allowed the main body of the commonplace book to have its own unruly, unplanned meanderings. Imposing too much order runs the risk of orphaning a promising hunch in a larger project that has died, and it makes it difficult for those ideas to mingle and breed when you revisit them. You need a system for capturing hunches, but not necessarily categorizing them, because categories can build barriers between disparate ideas, restrict them to their own conceptual islands. This is one way in which the human history of innovation deviates from the natural history. New ideas do not thrive on archipelagos.

In the bibliographic history of epic miscellany, another British title deserves mention alongside Erasmus Darwin's commonplace book: an immensely popular Victorian how-to guide with the memorable title *Enquire Within Upon Everything*. The frontispiece text for the book, first published in 1865, hints at the immense collection of domestic resources it contained:

Whether you wish to model a flower in wax; to study the rules of etiquette; to serve a relish for breakfast or supper; to supply a delicious entrée for the dinner table; to plan a dinner for a large party or a small one; to cure a headache; to make a will; to get married; to bury a relative; whatever you may wish to do, to make, or to enjoy, provided your desire has relation to the necessities of domestic life, I shall be happy to assist you and therefore hope you will not fail to Enquire Within—Editor.

Over a hundred editions of the guide were published, and it remained a common staple of British households well into the twentieth century. One musty copy of the book lingered into the 1960s, in the home of a pair of mathematicians living in the suburbs of London. The couple had a young son who was drawn to the "suggestion of magic" in the book's title, and who spent hours exploring this "portal to the world of information." The title stuck in the back of his mind, along with that wondrous feeling of exploring an immense trove of data. More than a decade later, he was working as a software consultant in a Swiss research lab, and found himself overwhelmed by the flow of information and the personnel churn in the organization. As a side project, he began tinkering with an application that would allow him to keep track of all that data. When it came time to give his program a name, his mind drifted back to that strange Victorian household encyclopedia from his youth. He called his application Enquire.

The application allowed you to store small blocks of information about people or projects as nodes in a connected network. It was easy to attach two-way pointers between nodes, so if you pulled up a person's name, you could instantly see all the projects he or she was working on. The application proved to be genuinely informative, but the programmer soon switched jobs and abandoned the code. He started up another version, which he called Tangle, a few years later, but it never got off the ground. But then, almost ten years after he had first programmed Enquire, he began sketching out a more ambitious application that could make connections between documents stored on different computers, using hypertext links. For a while he struggled with the right name for his nascent platform, calling it an information "mine" or "mesh." Eventually, he hit upon a different metaphor for the platform's dense network. He called it the World Wide Web.

In his own account of the Web's origins, Tim Berners-Lee makes no attempt to collapse the evolution of his marvelous idea into a single epiphany. The Web came into being as an archetypal slow hunch: from a child's exploration of a hundred-year-old encyclopedia, to a freelancer's idle side project designed to help him keep track of his colleagues, to a deliberate attempt to build a new information platform that could connect computers across the planet. Like Darwin's great understanding of life's tangled web, Berners-Lee's idea needed time—at least a decade's worth—to mature:

> Journalists have always asked me what the crucial idea was, or what the singular event was, that allowed the Web to exist one day when it hadn't the day before. They are frustrated when I tell them there was no "Eureka!" moment . . . Inventing the World Wide Web involved my growing realization that there was a power in arranging ideas in an unconstrained, weblike way.

And that awareness came to me through precisely that kind of process. The Web arose as the answer to an open challenge, through the swirling together of influences, ideas, and realizations from many sides, until, by the wondrous offices of the human mind, a new concept jelled. It was a process of accretion, not the linear solving of one problem after another.

Berners-Lee's slow, accretive development of the Web takes us to the next scale of innovation. Cultivating hunches extends beyond the private dominion of memory and the commonplace book. Most people do not have the luxury that Darwin had, of spending an entire life in pursuit of his intellectual fancies. For most people, ideas happen in and around their work environments, with all the daily pressures, distractions, accountability, and constant supervision that work life so often implies. In this respect, Berners-Lee was supremely lucky in the work environment he had settled into, the Swiss particle physics lab CERN. It took him ten years to nurture his slow hunch about a hypertext information platform. He spent most of those years working at CERN, but it wasn't until 1990—a decade after he had first begun working on Enquire—that CERN officially authorized him to work on the hypertext project. His day job was "data acquisition and control"; building a global communications platform was his hobby. Because the two shared some attributes, his superiors at CERN allowed Berners-Lee to tinker with his side project over the years. Thanks to a handful of newsgroups on the Internet, Berners-Lee was able to supplement and refine his ideas by conversing with other early hypertext innovators. That combination of flexibility and connection gave Berners-Lee critical support for his idea. He needed a work environment that carved out a space for slow hunches, cordoned off from all the immediate dictates of the day's agenda. And he needed information networks that let those hunches travel to other minds, where they could be augmented and polished.

If there is an innovation antimatter to CERN's hunch-sustaining matter, it might well be the Federal Bureau of Investigation in the summer of 2001. There were two crucial networks that failed to make the proper connections in the months leading up to 9/11: the information network of the Automated Case Support system, and the neural networks in the brains of the key participants. Even back in 2001, retrieving documents with an unlikely combination of terms—say, for example, "flight schools" and "radical Islamic fundamentalists"—was a routine matter; millions of users of Google, founded three years earlier, were doing comparable queries of the entire Web, with near-instantaneous results. Had the information network automatically suggested that the Radical Fundamentalist Unit officials read the Phoenix memo after the Minnesota office began its investigation into Moussaoui, the last few weeks of summer might have played out very differently. But however smart the network itself could have been, it still required a comparable connection to take place in the minds of the participants. If David Frasca had read the memo that Ken Williams had addressed to him, he might well have been able to connect the two hunches, using the advanced pattern recognition technology of the human brain.

The failure of those two networks to connect the Phoenix and Minnesota hunches was partly attributable to the practically medieval information technology employed by the FBI. But even if the Bureau had miraculously upgraded its network in the summer of 2001, the two hunches would likely have remained apart, because the lack of connections in the Automated Case Support system was a design principle, not merely the result of old-fashioned technology. It was, in computer-science parlance, a feature, not a bug. The FBI's information network was a classic closed network: not only could outsiders not access information in it, but also, the system was designed so that documents were carefully shielded from other members of the organization, a legacy of an institution predicated on secrets and "need to know" restrictions. The final report of the Judiciary Committee investigation into the intelligence failings in the months prior to 9/11 explicitly cited this design principle of the Bureau's information network as one of the key culprits, calling it "a 'stove pipe' mentality where crucial intelligence is pigeonholed into a particular unit and may not be shared with other units."

In a real sense, the FBI in the months leading up to 9/11 was a hunch-killing system, which is more than a little ironic, given the important role that hunches play in most accounts—real or fictional—of great investigators. In the FBI culture, an analyst labeling a report "speculative" was enough to keep it from advancing up the chain of command, while the outdated stovepipe architecture kept Williams's hunch from circulating to other field agents working on their own hunches. Tim Berners-Lee's monumental vision at CERN was a tangled web of data, a "swirling together of influences, ideas, and realizations from many sides." The Automated Case Support system wasn't just incompetent at creative tangle; the system was explicitly designed to eliminate it.

. . .

In 1980, to allude to *Enquire Within Upon Everything* in the name of your software package was more than a little audacious; Berners-Lee was just trying to keep track of his colleagues at CERN, not organize all the world's information. But "enquire within . . ." could well be the slogan for Google, which is why it is strangely appropriate that Google, in its own corporate environment, has arguably done the most to adopt and expand the kind of slow-hunch innovation that created the Web in the first place. Early in its history, Google famously instituted a "20-percent time" program for all Google engineers: for every four hours they spend working on official company projects, the engineers are *required* to spend one hour on their own pet project, guided entirely by their own passions and instincts. (Modeled on a similar program pioneered by 3M known as "the 15-percent rule," Google's system is officially called "Innovation Time Off.") The only requirements are that they give semiregular updates on their progress to their superiors. Most engineers end up drifting from idea to idea, and the vast majority of those ideas never turn into an official Google product. But every now and then, one of those hunches blooms into something significant. AdSense, Google's platform that allows bloggers and Web publishers to run Google ads on their sites, was partially generated during 20-percent time. In 2009, AdSense was responsible for more than $5 billion of Google's earnings, nearly a third of their total for the year. Orkut, one of the largest social network sites in India and Brazil, originated in the Innovation Time Off of a Turkish Google engineer named Orkut Büyükkökten. Google's popular mail platform, Gmail, has roots in an Innovation Time Off project as well. Marissa Mayer, Google's vice president of Search Products and User Experience, claims that over 50 percent of Google's new products derive from Innovation Time Off hunches.

The most telling contrast between Google and the FBI lies in the story of Krishna Bharat, who now holds the title of "principal scientist" at Google. In the weeks after 9/11, Bharat found himself overwhelmed by the amount of news information available about the attacks and the imminent war in Afghanistan. It occurred to him that it would be useful to create a software tool that could organize all those stories into useful clusters of relevance, so that you could see at a glance all the latest stories from around the Web about the search for bin Laden, or the cleanup efforts at Ground Zero, or the Bush administration's case for military retaliation. Bharat decided to use his 20-percent time to build a system called StoryRank—modeled after the original PageRank algorithm that Google's search engine relies on—to organize and cluster news items. StoryRank eventually blossomed into Google News, one of the most popular (and controversial) sources of news and commentary on the Web.

In a sense, the narrative of StoryRank's evolution is the exact mirror image of the narrative of the Phoenix memo. Like Tim Berners-Lee, Bharat was blessed with an organizational culture that encouraged hunches and gave them the space and time they needed to evolve. And Bharat took that nurturing environment and used it to build a tool that could automatically assemble clusters of relevance and association between documents—precisely the kind of system that could have connected the dots between the Phoenix memo and the Moussaoui investigation. Bharat had a hunch in his mind that there was a better way to organize the information net-work of news, and what he built turned out to be a tool that could be used to help related hunches complete one another.

Google News launched in September of 2002, which means StoryRank went from a hunch in Krishna Bharat's mind to a ship-ping product in one year. Nine years after 9/11, the FBI is still using the Automated Case Support system.

IV.

SERENDIPITY

Like any other thought, a hunch is simply a network of cells firing inside your brain in an organized pattern. But for that hunch to blossom into something more substantial, it has to connect with other ideas. The hunch requires an environment where surprising new connections can be forged: the neurons and synapses of the brain itself, and the larger cultural environment that the brain occupies.

For many years a debate raged over the nature of those neural connections: Were they chemical or electrical in nature? Were there chemical soups in the brain, or sparks? The answer turned out to be: both. Neurons send electrical signals down the long cables of their axons, which connect to other neurons via small synaptic gaps. When the electrical charge reaches the synapse, it releases a chemical messenger—a neurotransmitter, like dopamine or serotonin—that floats across to the receiving neuron and ultimately triggers another electrical charge, which travels out to other neurons in the brain.

The hybrid electrochemical nature of nerve communication was first established in another of the twentieth century's most celebrated experiments. In the early 1920s, the German scientist Otto Loewi isolated two still-beating frog hearts in separate vessels containing a saline solution. In one heart, he attached an electrode to the vagus nerve, which in an intact body starts in the brain stem and extends throughout the body. Because the vagus nerve helps regulate the parasympathetic system, stimulating the nerve with an electric charge slowed the heartbeat down. Loewi then extracted some of the solution that surrounded the heart and poured it over the second heart. Instantly, the second heart began to beat more slowly as well, even though its vagus nerve had not been electrically stimulated. Loewi's ingenious experiment demonstrated that the instructions to slow down the heartbeat had passed through the chemical soup of the saline solution. By stimulating a different part of the frog's vagus nerve, he could also accelerate both heartbeats in the same fashion. We now know that the electrical stimulation was releasing two distinct molecules into the soup: acetylcholine (which slowed the heart down) and adrenaline (which stimulated it).

Loewi's experiment, as influential as it was, is now remembered as much for the curious way Loewi conceived of it. The idea for the experiment came to Loewi in a dream—in two dreams, to be exact:

> The night before Easter Sunday of that year I awoke, turned on the light, and jotted down a few notes on a tiny slip of thin paper. Then I fell asleep again. It occurred to me at six o'clock in the morning that during the night I had written down something most important, but I was unable to decipher the scrawl. The next night, at three o'clock, the idea returned. It was the

design of an experiment to determine whether or not the hypothesis of chemical transmission that I had uttered seventeen years ago was correct. I got up immediately, went to the laboratory, and performed a simple experiment on a frog heart according to the nocturnal design.

We conventionally associate dream inspiration with the creative arts, but the canon of scientific breakthroughs contains many revolutionary ideas that originated in dreams. The Russian scientist Dmitri Mendeleev created the periodic table of the elements after a dream suggested to him that the table could be ordered by atomic weight. It was in a dream in 1947 that Nobel laureate John Carew Eccles originally conceived his theory of synaptic inhibitory action, which helped explain how connected neurons can fire without triggering an endless cascade of brain activity. Interestingly, Eccles's initial hunch involved a purely electrical system, but later experiments proved that the chemical GABA was central to synaptic inhibition, putting him in agreement with Loewi's experiment of decades before.

There is nothing mystical about the role of dreams in scientific discovery. While dream activity remains a fertile domain for research, we know that during REM sleep acetylcholine-releasing cells in the brain stem fire indiscriminately, sending surges of electricity billowing out across the brain. Memories and associations are triggered in a chaotic, semirandom fashion, creating the hallucinatory quality of dreams. Most of those new neuronal connections are meaningless, but every now and then the dreaming brain stumbles across a valuable link that has escaped waking consciousness. In this sense, Freud had it backward with his notion of dreamwork: the dream is not somehow unveiling a repressed truth. Instead, it is *exploring,* trying to find new truths by experimenting with novel combinations of neurons.

A recent experiment led by the German neuroscientist Ullrich Wagner demonstrates the potential for dream states to trigger new conceptual insights. In Wagner's experiment, test subjects were assigned a tedious mathematical task that involved the repetitive transformation of eight digits into a different number. With practice, the test subjects grew steadily more efficient at completing the task. But Wagner's puzzle had a hidden pattern to it, a rule that governed the numerical transformations. Once discovered, the pattern allowed the subjects to complete the test much faster, not unlike the surge of activity one gets at the end of a jigsaw puzzle when all the pieces suddenly fall into place. Wagner found that after an initial exposure to the numerical test, "sleeping on the problem" more than doubled the test subjects' ability to discover the hidden rule. The mental recombinations of sleep helped them explore the full range of solutions to the puzzle, detecting patterns that they had failed to perceive in their initial training period. The work of dreams turns out to be a particularly chaotic, yet productive, way of exploring the adjacent possible.

In a sense, dreams are the mind's primordial soup: the medium that facilitates the serendipitous collisions of creative insight. And hunches are like those early carbon atoms, seeking out new kinds of connections to help them build new chains and rings of innovation. Loewi's dream about the frog heart experiment is often invoked as a story of sudden epiphany—a twentieth-century version of Newton's apple—but the truth is that Loewi had been musing on the idea that nerves might communicate chemically for seventeen years. In part, his epiphany was made possible by the random connections of REM sleep. Yet it was also made possible by a slow hunch that had been lingering in the back of his mind for almost two decades.

This pattern of a slow hunch crystallizing into a dream-inspired epiphany recurs in what may be the most famous reverie in the history of science. In 1865, the German chemist Friedrich August Kekulé von Stradonitz had a daydream by a crackling fire in which he saw a vision of Ouroboros, the serpent from Greek mythology that devours its own tail. Kekulé had spent the past ten years of his life exploring the connections of carbon-based molecules. The serpent image in his dream gave him a sudden insight into the molecular structure of the hydrocarbon benzene. The benzene molecule, he realized, was a perfect ring of carbon, with hydrogen atoms surrounding its outer edges. Kekulé's slow hunch had set the stage for the insight, but for that hunch to turn into a world-changing idea, he needed the most unlikely of connections: an iconic image from ancient mythology. And Kekulé's vision did indeed prove to be a breakthrough of epic proportions: the ring structure of the benzene molecule became the basis for a revolution in organic chemistry, opening up a new vista onto the mesmerizing array of rings, lattices, and chains formed by that most connective of elements, carbon. It took the combinatorial serendipity of a daydream—all those neurons firing in unlikely new configurations—to help us understand the combinatorial power of carbon, which was itself crucial to understanding the original innovations of life itself.

The waking brain, too, has an appetite for the generative chaos that rules in the dream state. Neurons share information by passing chemicals across the synaptic gap that connects them, but

they also communicate via a more indirect channel: they synchro-
nize their firing rates. For reasons that are not entirely understood,
large clusters of neurons will regularly fire at the exact same
frequency. (Imagine a discordant jazz band, each member following
a different time signature and tempo, that suddenly snaps into a
waltz at precisely 120 beats per minute.) This is what neuroscien-
tists call phase-locking. There is a kind of beautiful synchrony to
phase-locking—millions of neurons pulsing in perfect rhythm. But
the brain also seems to require the opposite: regular periods of electrical chaos, where neurons are completely out of sync with each
other. If you follow the various frequencies of brain-wave activity
with an EEG, the effect is not unlike turning the dial on an AM
radio: periods of structured, rhythmic patterns, interrupted by static
and noise. The brain's systems are "tuned" for noise, but only in
controlled bursts.

In 2007, Robert Thatcher, a brain scientist at the University of
South Florida, decided to study the vacillation between phase-lock
and noise in the brains of dozens of children. While Thatcher found
that the noise periods lasted, on average, for 55 milliseconds, he also
detected statistically significant variation among the children. Some
brains had a tendency to remain longer in phase-lock, others had
noise intervals that regularly approached 60 milliseconds. When
Thatcher then compared the brain-wave results with the children's
IQ scores, he found a direct correlation between the two data sets.
Every extra millisecond spent in the chaotic mode added as much
as twenty IQ points. Longer spells in phase-lock *deducted* IQ points,
though not as dramatically.

Thatcher's study suggests a counterintuitive notion: the more disorganized your brain is, the smarter you are. It's counterintuitive in part because we tend to attribute the growing intelligence of the technology world with increasingly precise electromechanical choreography. Intel doesn't advertise its latest microprocessors with the slogan: "Every 55 milliseconds, our chips erupt into a blizzard of noise!" Yet somehow brains that seek out that noise seem to thrive, at least by the measure of the IQ test.

Science does not yet have a solid explanation for the brain's chaos states, but Thatcher and other researchers believe that the electric noise of the chaos mode allows the brain to experiment with new links between neurons that would otherwise fail to connect in more orderly settings. The phase-lock mode (the theory goes) is where the brain executes an established plan or habit. The chaos mode is where the brain assimilates new information, explores strategies for responding to a changed situation. In this sense, the chaos mode is a kind of background dreaming: a wash of noise that makes new connections possible. Even in our waking hours, it turns out, our brains gravitate toward the noise and chaos of dreams, 55 milliseconds at a time.

William James, writing in the late 1880s, had no way of measuring synchronized neuron firing, but his description of the "highest order of minds" captures something of the chaos mode:

> Instead of thoughts of concrete things patiently following one another, we have the most abrupt cross-cuts and transitions from one idea to another, the most rarefied abstractions and discriminations, the most unheard-of combinations of elements . . . a seething caldron of ideas, where everything is fizzling and bob-

bing about in a state of bewildering activity, where partnerships can be joined or loosened in an instant, treadmill routine is unknown, and the unexpected seems the only law.

The act of sexual reproduction is itself a kind of testament to the power of random connections, even in the most monogamous relationships. The overwhelming majority of nonmicroscopic life on earth produces offspring by sharing genes with another organism. But the evolution of this reproductive strategy remains something of a mystery. It would have been far easier for life to have avoided the complicated genetic exchanges of meiosis and fertilization. (Think of the elaborate system that the flowering plants had to evolve, luring insects to take on the task of carrying pollen from flower to flower.) Reproduction without sex is a simple matter of cloning; you take your own cells, make a copy, and pass that on to your descendants. It doesn't sound like much fun to our mammalian ears, but it's a strategy that has worked very well for billions of years for bacteria. Asexual reproduction is faster and more energy efficient than the sexual variety: you don't need to go to the trouble of finding a partner in order to create the next generation.

If natural selection rewarded organisms exclusively for sheer reproductive power, sexual reproduction might never have evolved. Asexual organisms reproduce on average twice as quickly as their sexual counterparts, in part because without a male/female distinction, every organism is capable of producing offspring directly. But evolution is not just a game of sheer quantity. Overpopulation, after all, poses its own dangers, and a community of organisms with identical DNA makes a prime target for parasites or predators. For

these reasons, natural selection also rewards innovation, life's tendency to discover new ecological niches, new sources of energy. This
is what Stuart Kauffman recognized when he first formulated the
idea of the adjacent possible: that there is something like an essential drive in the biosphere to diversify into new ways of making a
living. Scrambling together two distinct sets of DNA with each generation made for a far more complicated reproductive strategy, but
it paid immense dividends in the rate of innovation. What we gave
up in speed and simplicity, we made up for in creativity.

The water flea Daphnia lives in most freshwater ponds and
swamps. Its spasmodic movements in the water are responsible for
the "flea" description, but in reality Daphnia are tiny crustaceans,
no more than a few millimeters long. Under normal conditions,
Daphnia reproduce asexually, with females producing a brood of
identical copies of themselves in a tiny pouch. In this mode, the
Daphnia community is composed entirely of females. This reproductive strategy proves to be stunningly successful: in warm summer months, Daphnia will often be one of the most abundant
organisms in a pond ecosystem. But when conditions get tough,
when droughts or other ecological disturbances happen, or when
winter rolls in, the water fleas make a remarkable transformation:
they start producing males and switch to reproducing sexually. In
part, this switch is attributable to the sturdier eggs produced by
sexual reproduction, which are more capable of surviving the long
months of winter. But scientists believe that the sudden adoption
of sex is also a kind of biological innovation strategy: in challenging times, an organism needs new ideas to meet those new challenges. Reproducing asexually makes perfect sense during pros-
perous periods: if life is good, keep doing what you're doing. Don't

mess with success by introducing new genetic combinations. But when the world gets more challenging—scarce resources, predators, parasites—you need to innovate. And the quickest path to innovation lies in making novel connections. This strategy of switching back and forth between asexual and sexual reproduction goes by the name "heterogamy," and while it is unusual, many different organisms have adopted it. Slime molds, algae, and aphids have all evolved heterogamous reproductive strategies. In each organism, the Daphnia pattern repeats itself: the genetic recombinations of sex emerge when conditions get difficult. Swapping genes with another organism is itself more difficult than simple cloning, but the innovation rewards of sex outweigh the risks of the more stable path. When nature finds itself in need of new ideas, it strives to connect, not protect.

The English language is blessed with a wonderful word that captures the power of accidental connection: "serendipity." First coined in a letter written by the English novelist Horace Walpole in 1754, the word derives from a Persian fairy tale titled "The Three Princes of Serendip," the protagonists of which were "always making discoveries, by accident and sagacity, of things they were not in quest of." The contemporary novelist John Barth describes it in nautical terms: "You don't reach Serendip by plotting a course for it. You have to set out in good faith for elsewhere and lose your bearings serendipitously."

But serendipity is not just about embracing random encounters for the sheer exhilaration of it. Serendipity is built out of happy accidents, to be sure, but what makes them happy is the fact that the discovery you've made is meaningful to you. It completes a hunch, or opens up a door in the adjacent possible that you had overlooked. If you're a geologist randomly exploring the Web, and the particular isle of Serendip that you stumble across turns out to be an essay on health-care reform, your discovery may well be interesting and informative, but it will not be truly serendipitous unless it helps you fill in a piece of a puzzle you've been poring over. That's not to say that geologists can only find serendipitous discoveries in texts about geology—quite the contrary, in fact. Serendipitous discoveries often involve exchanges across traditional disciplines. Think of the way Kekulé's mythic serpent led to a revolution in organic chemistry. It was genuinely serendipitous that Kekulé's dreaming brain should conjure up the image of Ouroboros at that moment. But had Kekulé not been wrestling with the structure of the benzene molecule for years, that serpent shape might not have triggered any useful associations in his mind. (Sometimes a serpent swallowing its tail is just a serpent swallowing its tail, as Freud might have said.) Serendipity needs unlikely collisions and discoveries, but it also needs something to anchor those discoveries. Otherwise, your ideas are like carbon atoms randomly colliding with other atoms in the primordial soup without ever forming the rings and lattices of organic life.

The challenge, of course, is how to create environments that foster these serendipitous connections, on all the appropriate scales: in the private space of your own mind; within larger institutions; and across the information networks of society itself.

At first blush, the idea of conjuring up serendipitous discoveries inside your own mind seems like a contradiction in terms. Wouldn't that be like losing your bearings in your own driveway? Yet that's exactly what Kekulé was doing by the fire. He was connecting two distinct thoughts that each occupied a slot in his memory banks: the riddle of benzene's molecular structure, and the tail-swallowing Ouroboros. The truth is, your mind contains a near-infinite number of ideas and memories that at any given moment are lurking outside your consciousness. Some tiny fraction of those thoughts are like Kekulé's serpent: surprising connections that might help you unlock a door in the adjacent possible. But how do you get those particular clusters of neurons to fire at the right time?

One way is to go for a walk. The history of innovation is replete with stories of good ideas that occurred to people while they were out on a stroll. (A similar phenomenon occurs with long showers or soaks in a tub; in fact, the original "eureka" moment—Archimedes hitting upon a way of measuring the volume of irregular shapes—occurred in a bathtub.) The shower or stroll removes you from the task-based focus of modern life—paying bills, answering e-mail, helping kids with homework—and deposits you in a more associative state. Given enough time, your mind will often stumble across some old connection that it had long overlooked, and you experience that delightful feeling of private serendipity: Why didn't I think of that before?

In his book *The Foundations of Science*, the French mathematician and physicist Henri Poincaré devotes an autobiographical chapter to the question of mathematical creativity. The chapter begins with a detailed account of how Poincaré discovered the class of Fuchsian functions, one of the first influential mathematical concepts of his career. He begins by attempting to prove that the functions do not exist; for fifteen days he struggles at his desk with no success. Then one evening he breaks from his ordinary routine and drinks black coffee. Unable to sleep, his mind seethes with promising hunches. "Ideas rose in crowds," Poincaré writes. "I felt them collide until pairs interlocked, so to speak, making a stable combination. By the next morning I had established the existence of a class of Fuchsian functions, those which come from the hypergeometric series." His next insight—a connection between the functions and non-Euclidean geometry—comes several weeks later, while boarding a bus during a geological expedition in Normandy. On his return home, he commences work on an unrelated arithmetical question and flounders for several days. "Disgusted with my failure," he writes, "I went to spend a few days at the seaside, and thought of something else. One morning, walking on the bluff, the idea came to me, with just the same characteristics of brevity, suddenness and immediate certainty, that the arithmetic transformations of indeterminate ternary quadratic forms were identical with those of non-Euclidean geometry." He returns home again and works through the implications, but encounters another roadblock. Military service then dictates a trip to Fort Mont-Valérien in the suburbs of Paris, where he has little time to think about mathematics at all. And yet the final missing piece arrives nonetheless. "One day, going along the street, the solution of the difficulty which had stopped me suddenly appeared to me. I did not try to go deep into it immediately, and only after my service did I again take up the question. I had all the elements and had only to arrange them and put them together. So I wrote out my final memoir at a single stroke and without difficulty."

Poincaré's account may be the most "pedestrian" story of scientific creativity on record. Whenever he actually sits down at his desk, the innovations seem to grind to a halt. But on foot, his ideas "rose in crowds." Trying to explain the phenomenon, Poincaré reaches for an atomic metaphor, with each partial idea or hunch represented by an atom hooked to a wall. In normal situations, the atoms remain in place, locked into a stable configuration. But when the mind wanders (and, in Poincaré's case, when the physical body wanders), the atoms become untethered. "During a period of apparent rest and unconscious work, certain of them are detached from the wall and put in motion. They flash in every direction through the space . . . where they are enclosed, as would, for example, a swarm of gnats or, if you prefer a more learned comparison, like the molecules of gas in the kinematic theory of gases. Then their mutual impacts may produce new combinations."

While the creative walk can produce new serendipitous combinations of existing ideas in our heads, we can also cultivate serendipity in the way that we absorb new ideas from the outside world. Reading remains an unsurpassed vehicle for the transmission of interesting new ideas and perspectives. But those of us who aren't scholars or involved in the publishing business are only able to block out time to read around the edges of our work schedule: listening to an audio book during the morning commute, or taking in a chapter after the kids are down. The problem with assimilating new ideas at the fringes of your daily routine is that the potential combinations are limited by the reach of your memory. If it takes you two weeks to finish a book, by the time you get to the next book, you've forgotten much of what was so interesting or provocative about the original one. You can immerse yourself in a single author's perspective, but then it's harder to create serendipitous collisions between the ideas of multiple authors. One way around this limitation is to carve out dedicated periods where you read a large and varied collection of books and essays in a condensed amount of time. Bill Gates (and his successor at Microsoft, Ray Ozzie) are famous for taking annual reading vacations. During the year they deliberately cultivate a stack of reading material—much of it unrelated to their day-to-day focus at Microsoft—and then they take off for a week or two and do a deep dive into the words they've stockpiled. By compressing their intake into a matter of days, they give new ideas additional opportunities to network among themselves, for the simple reason that it's easier to remember something that you read yesterday than it is to remember something you read six months ago.

In Poincaré's language, the deep dive, like the long stroll, detaches the atoms from the wall and puts them in motion. Most of us don't have the luxury of taking deep dive reading sabbaticals, of course, and reading a few thousand pages is not everyone's idea of a fun vacation. But there's no reason why organizations couldn't recognize the value of a reading sabbatical, the way many organizations encourage their employees to take time off for learning new skills. If Google can give its engineers one day a week to work on anything they want, surely other organizations can figure out a way to give their employees dedicated time to immerse themselves in a network of new ideas.

Private serendipity can be cultivated by technology as well. For more than a decade now, I have been curating a private digital archive of quotes that I've found intriguing, my twenty-first-century version of the commonplace book. Some of these passages involve very focused research on a specific project; others are more random discoveries, hunches waiting to make a connection. Some of them are passages that I've transcribed from books or articles; others were clipped directly from Web pages. (In the past few years, thanks to Google Books and the Kindle, copying and storing interesting quotes from a book has grown far simpler.) I keep all these quotes in a database using a program called DEVONthink, where I also store my own writing: chapters, essays, blog posts, notes. By combining my own words with passages from other sources, the collection becomes something more than just a file storage system. It becomes a digital extension of my imperfect memory, an archive of all my old ideas, and the ideas that have influenced me. There are now more than five thousand distinct entries in that database, and more than 3 million words—sixty books' worth of quotes, fragments, and hunches, all individually captured by me, stored in a single database.

Having all that information available at my fingertips is not just a quantitative matter of finding my notes faster. Yes, when I'm trying to track down an article I wrote many years ago, it's now much easier to retrieve. But the qualitative change lies elsewhere: in finding documents that I've forgotten about altogether, finding documents that I didn't know I was looking for. What makes the system truly powerful is the way that it fosters private serendipity.

DEVONthink features a clever algorithm that detects subtle semantic connections between distinct passages of text. These tools are smart enough to get around the classic search-engine failing of excessive specificity: searching for "dog" and missing all the articles that only have the word "canine" in them. Modern indexing software like DEVONthink's learns associations between individual words by tracking the frequency with which words appear near each other. This can create almost lyrical connections between ideas. Several years ago, I was working on a book about cholera in London and queried DEVONthink for information about Victorian sewage systems. Because the software had detected that the word "waste" is often used alongside "sewage," it directed me to a quote that explained the way bones evolved in vertebrate bodies: namely, by repurposing the calcium waste products created by the metabolism of cells. At first glance that might seem like an errant result, but it sent me off on a long and fruitful tangent into the way complex systems—whether cities or bodies—find productive uses for the waste they create. That idea became a central organizing theme for one of the chapters in the cholera book. (It will, in fact, reappear in this book in a different guise.)

Now, strictly speaking, who was responsible for that initial idea? Was it me, or the software? It sounds like a facetious question, but I mean it seriously. Obviously, the computer wasn't *conscious* of the idea taking shape, and I supplied the conceptual glue that linked the London sewers to cell metabolism. But I'm not at all confident that I would have made the initial connection without the help of the software. The idea was a true collaboration, two very different kinds of intelligence playing off one another, one carbon-based, the other silicon. When I'd first captured that quote about calcium and bone structure, I'd had no idea that it would ultimately connect to the history of London's sewage system (or to a book about innovation). But there was something about that concept that intrigued me enough to store it in the database. It lingered there for years in the software's primordial soup, a slow hunch waiting for its connection.

I use DEVONthink as an improvisational tool as well. I write a paragraph about something—let's say it's about the human brain's remarkable facility for interpreting facial expressions. I then plug that paragraph into the software, and ask DEVONthink to find other passages in my archive that are similar. Instantly, a list of quotes appears on my screen: some delving into the neural architecture that triggers facial expressions, others exploring the evolutionary history of the smile, others dealing with the expressiveness of our near-relatives, the chimpanzees. Invariably, one or two of these triggers a new association in my head—perhaps I've forgotten about the chimpanzee connection—and so I select that quote, and ask the software to find a new batch of passages similar to it. Before long, a larger idea takes shape in my head, built upon the trail of associations the machine has assembled for me.

Compare that to the traditional way of exploring your files, where the computer is like a dutiful, but dumb, butler: "Find me that document about the chimpanzees!" That's searching. The other feels radically different, so different that we don't quite have a verb for it: it's riffing, or exploring. There are false starts and red herrings, but there are just as many happy accidents and unexpected discoveries. Indeed, the fuzziness of the results is part of what makes the software so powerful. The serendipity of the system emerges out of two distinct forces. First, there is the connective power of the semantic algorithm, which is smart but also slightly unpredictable, thus creating a small amount of randomizing noise that makes the results more surprising. But that randomizing force is held in check by the fact that I have curated all these passages myself, which makes each individual connection far more likely to be useful to me in some way. When you start a new query in DEVONthink and look down at the initial results, at first glance they can sometimes seem jumbled and disconnected, but then you read through them in more detail, and inevitably something tantalizing catches your eye. "Jumbled" and "disconnected" is of course also how we describe the strange explorations of our dreams, and the comparison is an apt one. DEVONthink takes the strange but generative combinations of the dream state and turns them into software.

If you visit the "serendipity" entry in Wikipedia, you are one click away from entries on LSD, Teflon, Parkinson's disease, Sri Lanka, Isaac Newton, Viagra, and about two hundred other topics of comparable diversity. That eclecticism is particularly acute at Wikipedia, of course, but it derives from the fundamentally "tangled" nature of Tim Berners-Lee's original hypertext architecture. No medium in history has ever offered such unlikely trails of connection and chance in such an intuitive and accessible form. Yet in recent years, a puzzling meme has emerged on op-ed pages with a strange insistence: the rise of the Web, its proponents argue, has led to a *decline* in serendipitous discovery. Consider this representative elegy to the "endangered joy of serendipity," authored by a journalism professor named William McKeen:

> Think about the library. Do people browse anymore? We have become such a directed people. We can target what we want, thanks to the Internet. Put a couple of key words into a search engine and you find—with an irritating hit or miss here and there—exactly what you're looking for. It's efficient, but dull. You miss the time-consuming but enriching act of looking through shelves, of pulling down a book because the title interests you, or the binding . . . Looking for something and being surprised by what you find—even if it's not what you set out looking for—is one of life's great pleasures, and so far no software exists that can duplicate that experience.

In a similar piece, the *New York Times* technology editor, Damon Darlin, complained that the "digital age is stamping out serendipity." Darlin acknowledged the vast influx of suggested reading that now arrives on our screen every morning via social network services like Twitter and Facebook, but claimed those links didn't constitute serendipity. "[They're] really group-think," Darlin argued. "Everything we need to know comes filtered and vetted. We are discovering what everyone else is learning, and usually from people we have selected because they share our tastes."

When critics complain about the decline of serendipity, they habitually point to two "old media" mechanisms that allegedly have no direct equivalent on the Web. McKeen mentions the first one: browsing the stacks in a library (or a bookstore), "pulling down a book because the title interests you, or the binding." Old-style browsing does indeed lead to unplanned discoveries. But thanks to the connective nature of hypertext, and the blogosphere's exploratory hunger for finding new stuff, it is far easier to sit down in front of your browser and stumble across something completely brilliant but surprising than it is to walk through a library, looking at the spines of books. Does everyone use the Web this way? Of course not. But it is much more of a mainstream pursuit than randomly exploring the library stacks, pulling down books because you like the binding, ever was. This is the irony of the serendipity debate: the thing that is being mourned has actually gone from a fringe experience to the mainstream of the culture.

The second analog-era mechanism that encourages serendipity involves the physical limitations of the print newspaper, which forces you to pass by a collection of artfully curated stories on a variety of topics, before you open up the section that most closely matches your existing passions and knowledge. The legal scholar Cass Sunstein refers to this as an example of the "architecture of serendipity." On the way to the sports section or the comics or the business page, you happen to collide with a story about the abuses of African diamond mines, and something in the headline catches your eye. A thousand words later, you've learned something powerful about people living halfway around the world whose existence you had never contemplated before. And perhaps there is some kind of serendipitous click in that collision: you'd been looking for a new charitable cause to support, or contemplating buying your spouse a diamond ring. And then this story drops in your lap, and helps you complete the thought. You weren't looking for a story about diamond mines, but it was exactly what you needed.

This is indeed a superb example of serendipity, and there is no doubt that newspapers facilitated comparable accidental discoveries countless times over countless breakfast tables during their heyday. The question is whether the transition to the Web makes this sort of discovery more or less frequent. If you compare the front pages of the print and online versions of a newspaper, the Web actually appears to have the upper hand. The Internet scholar Ethan Zuckerman compared the front page of the *New York Times* with that of its Web cousin and found that the print version had twenty-three references on its front page to articles in the paper (either in the form of lead articles themselves, or short summaries teased below the fold). The front page at NYTimes.com, in Zuckerman's study, contained 315 links to articles and other forms of content. If the architecture of serendipity lies in stumbling across surprising connections while scanning the front page, then the Web is more than ten times as serendipitous as the classic print newspaper.

Sunstein would no doubt argue that many people bypass the front door of their online newspaper, going directly to the sports or business-section page that they've bookmarked, or to some other filter tailor-made to their preexisting interests. No doubt millions of people use comparable filters every morning. One could reasonably question whether people like this who have gone out of their way to avoid encountering the "big picture" of the newspaper front page were ever likely to stumble across the diamond-mining story at the breakfast table with a print paper, or ramble through the stacks of their local library. But Sunstein and Darlin and McKeen are indeed correct when they argue that the Internet gives us topical filters that were unheard of in the days of mass media. But those filters are only part of the story. Filters reduce serendipity (unless your particular interest lies in being surprised, which is part of the appeal of beautifully miscellaneous blogs like Boing Boing). But beyond bookmarking, filters are a second-generation addition to the architecture of the Web. They are not native to it. What *is* native to the Web's architecture are two key features that have been great supporters of serendipity: a global, distributed medium in which anyone can be a publisher, and a hypertext document structure in which it is trivial to jump from a newspaper article to an academic essay to an encyclopedia entry in a matter of seconds. The information diversity of the Web ensures that there is an endless supply of surprising information to stumble across, and the links of hypertext ensure that we can get to that information at lightning speed, or follow trails of improvised association that would have been painfully slow to follow in the age of print media. Ironically, the problem with the Web is that there's too much noise, too much chaos—that's why the filters were invented in the first place. We have filters because the Web has unleashed too much diversity and surprise, not because we have too little.

I happen to believe that the Web, as a medium, has pushed the culture toward more serendipitious encounters. The simple fact that information "browsing" and "surfing" are now mainstream pursuits makes a strong case for a rise in serendipity, compared to cultures dominated by books or mass media. But whether or not you accept the premise that the average media consumer experiences more serendipitous discoveries thanks to the Web, there can be little doubt that the Web is an unrivaled medium for serendipity if you are actively seeking it out. If you want to build a daily reading list of eclectic and diverse perspectives, you can stitch one together in your RSS reader or your bookmarks bar in a matter of minutes, for no cost, while sitting on your couch. Just as important, you can use the Web to fill out the context when you do stumble across some interesting new topic. The great oracle of the digital age, Google, is often invoked as a serendipity killer, because search queries function as a kind of on-demand filter that eliminates the 99.999 percent of the Web that is not relevant to the searcher's current interest. But when critics put Google on the side of filters, they assume that most queries are variations on the theme: "I'm passionately interested in x and would like to learn more about it." No doubt some unthinkably large number of Google users enter queries that take that basic shape every day. But there's another type of query that is just as valuable: "Someone just told me about x and I know nothing about it, but it sounds interesting. Tell me more." This is the subtle way in which Google supports the serendipitous aspects of the Web. Yes, it's true that by the time you've entered something into the Google search box, you're already invested in the topic. (This is why Web pioneer John Battelle calls it the "database of intentions.") But often that investment is directly correlated with your ignorance about the topic at hand: someone mentions in passing the poetry of John Ashbery, or the television show *Arrested Development*, or the tail-swallowing Ouroboros, and you think: "What's the deal with that? It sounds really interesting." Imagine it's 1980 and you're sitting at your breakfast table, reading the morning paper, and on the way to the sports page you stumble across an article on the front page about this provocative new idea of global warming that you've not yet encountered. You can read the article, to be sure, but when the article leaves you hankering for more information and context, where do you go? Turn on the television and hope that one of the three networks or PBS is running a news item or a documentary on the topic at that exact second? Get in the car, drive fifteen minutes down to your public library and check out a book on the subject? Go through all the magazines in your house, scouring their table-of-contents pages for any climate-change-related articles?

Let's say you live in a particularly information-rich household for the standards of 1980, and you happen to have a copy of the *Encyclopædia Britannica*. But of course the version you bought is actually the 1976 edition, and global warming doesn't make it into the *Britannica* until 1994, despite the fact that the term is common enough to be mentioned in ordinary parlance throughout the nineties.

Today, of course, you would query either Google or Wikipedia for the search term "global warming." And you would instantly have more information (and more perspectives) at your fingertips than would have been imaginable when you were thumbing through the *Britannica* in 1980. Yes, these results are targeted to your expressed interest in a specific topic, but that interest is often something you've just stumbled across, a hint more than a passion. And because those pages are built out of hyperlinks, just a few clicks can land you in an entirely new region of interest that you'd never dreamed of visiting. Google and Wikipedia give those passing hints something to attach to, a kind of information anchor that lets you settle down around a topic and explore the surrounding area. They turn hints and happy accidents into information. If the commonplace book tradition tells us that the best way to nurture hunches is to write everything down, the serendipity engine of the Web suggests a parallel directive: *look everything up*.

The premise that innovation prospers when ideas can serendipitously connect and recombine with other ideas, when hunches can stumble across other hunches that successfully fill in their blanks, may seem like an obvious truth, but the strange fact is that a great deal of the past two centuries of legal and folk wisdom about innovation has pursued the exact opposite argument, building walls between ideas, keeping them from the kind of random, serendipitous connections that exist in dreams and in the organic compounds of life. Ironically, those walls have been erected with the explicit aim of encouraging innovation. They go by many names: patents, digital rights management, intellectual property, trade secrets, proprietary technology. But they share a founding assumption: that in the long run, innovation will increase if you put restrictions on the spread of new ideas, because those restrictions will allow the creators to collect large financial rewards from their inventions. And those rewards will then attract other innovators to follow in their path.[4]

The problem with these closed environments is that they inhibit serendipity and reduce the overall network of minds that can potentially engage with a problem. This is why a growing number of large organizations—businesses, nonprofits, schools, government agencies—have begun experimenting with work environments that encourage the architecture of serendipity. Traditionally, organizations that have a strong demand for innovation have created a kind of closed playpen for hunches: the research-and-development lab. Ironically, R&D labs have historically functioned as a kind of idea lockbox; the hunches evolving in those labs tended to be the most heavily guarded secrets in the entire organization. Allowing these early product ideas to circulate more widely would allow rival firms to copy or exploit them. Some organizations—including Apple— have gone to great length to keep R&D experiments sequestered from other employees *inside* the organization.

But that secrecy, as we have seen, comes with great cost. Protecting ideas from copycats and competitors also protects them from other ideas that might improve them, might transform them from hints and hunches to true innovations. And indeed there is a grow-

---

4. Patents actually have a complicated historical relationship to the idea of open information networks. While most patent law is exclusive in nature—forbidding non-patent-holders from using a patented "method" without permission for a finite time period—patent law also conventionally involves an element of disclosure, where the inventor is forced to reveal the nature of his or her creation in technical detail. The disclosure is obviously partly designed to help enforce the restrictions in cases of patent infringement, but it was also intended to encourage good ideas to spread more freely, by making them part of the public record. Unfortunately, the modern emergence of patent trolls and squatters, supported by overzealous intellectual property lawyers, means that the protective side of patent law has dominated the connective side.

ing movement in some forward-thinking companies to turn their R&D labs inside out and make them far more transparent than the traditional model. Organizations like IBM and Procter & Gamble, who have a long history of profiting from patented, closed-door innovations, have embraced open innovation platforms over the past decade, sharing their leading-edge research with universities, partners, suppliers, and customers.

In early 2010, Nike announced a new Web-based marketplace it called the GreenXchange, where it publicly released more than 400 of its patents that involve environmentally friendly materials or technologies. The marketplace was a kind of hybrid of commercial self-interest and civic good. By making its good ideas public, Nike made it possible for outside firms to improve on those innovations, creating new value that Nike itself might ultimately be able to put to use in its own products. In a sense, Nike was widening the network of minds who were actively thinking about how to make its ideas more useful, without putting anyone else on its payroll. But Nike's organizational values also include a commitment to environmental sustainability, and the company recognized that many of its eco-friendly patents might be useful in different contexts. Nike is a big corporation, with many products in many categories, but there are limits to its reach. Some of its innovations might well turn out to be advantageous to industries or markets where it has no competitive involvement whatsoever. By keeping its eco-friendly ideas behind a veil of secrecy, Nike was holding back—without any real commercial justification—ideas that might, in another context, contribute to a sustainable future. In collaboration with Creative Commons, Nike released its patents under a modified license permitting use in "non-competitive" fields. (They also created a standardized, prenegotiated contract for the patents, thereby reducing the transaction costs of haggling over each patent license individually.) The example scenario they invoked at the launch of GreenXchange would have warmed the heart of Stephen Jay Gould: an environmentally sound rubber originally invented for use in running shoes that could be adapted by a mountain bike company to create more sustainable tires. Apparently, Gould's tires-to-sandals principle works both ways. Sometimes you make footwear by putting tires to new use, sometimes you make tires by putting footwear to new use. Green Xchange is trying to give multinational corporations some of the same freedom to reinvent and recycle that Gould's sandal-makers enjoy sifting through the Nairobi junkyards.

The other organizational technique for facilitating serendipitous connections is the "brainstorm" session, an approach pioneered by the advertising executive Alex Osborn in the 1930s. Brainstorming opens up the flow of ideas and hunches in a more generative fashion than is customary in a regimented workplace meeting. Yet a number of recent studies have suggested that brainstorming is less effective than its practitioners would like. One trouble with brainstorming is that it is finite in both time and space: a group gathers for an hour in a room, or for a daylong corporate retreat, they toss out a bunch of crazy ideas, and then the meeting disperses. Sometimes a useful connection emerges, but too often the relevant hunches aren't in sync with one another. One employee has a promising hunch in one office, and two months later, another employee comes up with the missing piece that turns that hunch into a genuine insight. Brainstorming might bring those two fragments together, but the odds are against it. Imagine some kind of alternate reality where the FBI holds a corporate retreat in late August of

2001, and invites the field agents from Arizona and Minnesota to sit in a room together and brainstorm new potential threats against the United States. No doubt it would have been the first corporate retreat on record that actually changed the fate of world history, but with more than ten thousand field agents across the nation, the odds against getting the right people from Arizona and Minnesota together at the right time would have been astronomical. But imagine if the FBI had been using a networked version of a DEVONthink archive instead of the archaic Automated Case Support system. The top brass at the Radical Fundamentalist Unit would still have read the search warrant request for Moussaoui's laptop and thought to themselves, "This sounds like a pretty shaky hunch." But a quick DEVONthink query would have pointed them to the Phoenix memo, to another hunch about flight training and terrorism. Those two unlikely ideas would have collided, without the field agents in Phoenix and Minnesota even speaking to each other, much less sitting down for a brainstorming session.

The secret to organizational inspiration is to build information networks that allow hunches to persist and disperse and recombine. Instead of cloistering your hunches in brainstorm sessions or R&D labs, create an environment where brainstorming is something that is constantly running in the background, throughout the organization, a collective version of the 20-percent-time concept that proved so successful for Google and 3M. One way to do this is to create an open database of hunches, the Web 2.0 version of the traditional suggestion box. A public hunch database makes every passing idea visible to everyone else in the organization, not just management. Other employees can comment or expand on those ideas, connecting them with their own hunches about new products or priorities or

internal organizational changes. Some systems even allow employees to vote on their colleagues' suggestions, not unlike the user
rankings that power collective news sites like Digg or Reddit.
Google has a company-wide e-mail list where employees can suggest new features or products; each suggestion can then be rated on
a scale of 0 ("Dangerous or harmful") to 5 ("Great idea! Make it
so."). Salesforce.com maintains a popular Idea Exchange where its
customers can suggest new features for the company's software
products. The Idea Exchange doesn't just allow interesting hunches
to circulate and connect. It also tracks their maturation into ship-
ping code: the front door of the Exchange includes prominent links
to submitted ideas currently being considered for inclusion in future releases, as well as ideas that were successfully integrated into
past releases. Too often, real-world suggestion boxes feel like a black
hole; you drop your idea in the slot, and never hear about it again.
In a public forum like Idea Exchange, not only do you get to see
and improve other people's suggestions, but you get tangible evidence that your ideas can make a difference.

These kinds of information networks can do a masterful job
of tapping both individual and collective intelligence: the individual employee has a provocative and useful hunch, and the group
helps complete the hunch by connecting it to other ideas that have
circulated through the system, and helps separate out that hunch
from the thousands of other less useful ones by voting it to the top
of the charts. By making the ideas public, and by ensuring that they
remain stored in the database, these systems create an architecture
for organizational serendipity. They give good ideas new ways to
connect.

v.

ERROR

In the summer of 1900 a twenty-seven-year-old aspiring inventor named Lee de Forest moved to Chicago, rented a one-room apartment on Washington Boulevard, and took a day job translating foreign articles on wireless technology for *Western Electrician* magazine. The translation work was informative: a major exposition on wireless technology that had just been held in Paris guaranteed a constant flow of interesting new research papers across the Atlantic. But de Forest's true passion lay in the cabinet of wonders he had assembled in his bedroom on Washington Boulevard: batteries, spark gap transmitters, electrodes—all the building blocks that would be assembled in the coming decade to invent the age of electronics.

For a budding innovator in wireless telegraphy at the turn of the century, the spark gap transmitter was the most essential of gadgets. Hertz and Marconi's original explorations of the electromagnetic spectrum had relied on spark gaps. The device employed two electrodes separated by a small gap. A battery attached to the electrodes supplied a surge of electricity, which caused a spark to jump from one electrode to the other, triggering a pulse of electromagnetic activity that could be detected and amplified by antennae miles away. Spark gap machines emitted a terse blast of monotone noise, perfect for sending Morse code.

On the night of September 10, 1900, de Forest was experimenting with his spark gap machine in the corner of his Washington Boulevard bedroom. Across the room, the red flame of a Welsbach burner flickered fifteen feet away. De Forest triggered a surge of voltage through the spark gap, and as the machine crackled, he could see the flame of the burner instantly change from red to white heat. De Forest later estimated that the flame's intensity had increased by several candlepower. Somehow, for reasons that de Forest could not explain, the electromagnetic pulse of the spark gap was intensifying the energy of a flame fifteen feet away. Watching that flame shift from red to white planted the seed of an idea in de Forest's head: that a gas could be employed as a wireless detector, one that might be more sensitive than anything Marconi or Tesla had created to date.

De Forest had stumbled across a classic slow hunch. In his autobiography, de Forest described the gas-flame detector as "a subject that had ever since been in the back of my mind." In the end, that hunch would mature into an invention that ultimately changed the landscape of the twentieth century, an invention that made radio, television, and the first digital computers possible. In 1903, he began a series of failed experiments with placing two electrodes in gas-filled glass bulbs. He continued tinkering with the model, until, several years later, he hit upon the idea of placing a third electrode in the bulb, attached to an antenna or external tuner. After a number of iterations, he used a piece of wire that had been bent back and forth several times as the middle electrode; de Forest called it the grid. Early tests showed that the device, which de Forest dubbed the Audion, proved far superior to other technology at amplifying audio signals without degrading the tuner's ability to separate out signals at different frequencies.

De Forest's creation would eventually be called a triode. Its three-electrode architecture would form the basis of the vacuum tubes that began to be mass-produced in the following decade. Radio receivers, telephone switchboards, television sets—all the communications revolutions of the first half of the century relied on some variation of de Forest's design to boost their signals. Initially employed for amplification, the vacuum tube turned out to have an unforeseen use as an electronic switch, enabling the high-speed logic gates of the first digital computers in the 1940s. When de Forest twisted the wire into the shape of a grid and placed it between those two electrodes, he was unwittingly opening up the adjacent possible for the Analytical Engine that Charles Babbage had failed to produce sixty years before. The power of that new portal was apparent instantly: the first computer built with vacuum tubes, the mammoth ENIAC, ran calculations that helped develop the hydrogen bomb.

The invention of the Audion sounds like a classic story of ingenuity and persistence: a maverick inventor holed up in his bedroom lab notices a striking pattern and tinkers with it for years as a slow hunch, until he hits upon a contraption that changes the world. But telling the story that way misses one crucial fact: that at almost every step of the way, de Forest was flat-out wrong about what he was inventing. The Audion was not so much an invention as it was the steady, persistent accumulation of error. The strange communication between the spark gap transmitter and the Wersbach gas burner flame turned out to have nothing to do with the electromagnetic spectrum. (The flame was responding to ordinary sound waves emitted by the spark gap transmitter.) But because de Forest had begun with this erroneous notion that the gas flame was detecting the radio signals, all his iterations of the Audion involved some low-pressure gas inside the device, which severely limited their reliability. It took another decade for researchers at General Electric and other firms to realize that the triode performed far more effectively in a true vacuum. (Hence the term "vacuum tube.") Even de Forest himself willingly admitted that he didn't understand the device he had invented. "I didn't know why it worked," he remarked. "It just did."

De Forest may have been the most erratic of the twentieth century's great inventors, but the error-prone history of his greatest success is hardly anomalous. The history of being spectacularly right has a shadow history lurking behind it: a much longer history of being spectacularly wrong, again and again. And not just wrong, but *messy*. A shockingly large number of transformative ideas in the annals of science can be attributed to contaminated laboratory environments. Alexander Fleming famously discovered the medical virtues of penicillin when the mold accidentally infiltrated a culture of *Staphylococcus* he had left by an open window in his lab. In the 1830s, Louis Daguerre spent years trying to coax images out of iodized silver plates. One night, after another futile attempt, he stored the plates in a cabinet packed with chemicals; to his wonder the next morning, the fumes from a spilled jar of mercury produced a perfect image on the plate—and the daguerreotype, forerunner of modern photography, was born.

In the summer of 1951, a World War II Navy veteran named Wilson Greatbatch was working at an animal behavior farm affiliated with the psychology department at Cornell, where he was studying under the G.I. Bill. Greatbatch had long been a ham radio enthusiast; as a teenager, he had built his own shortwave radio by cobbling together the descendants of de Forest's Audion. His love of gadgets had drawn him to the Cornell farm because the psychology department needed someone to attach experimental instruments to the animals, measuring their brain waves, heartbeats, and blood pressure. One day, Greatbatch happened to sit at lunch with two visiting surgeons and got into a conversation about the dangers of irregular heartbeats. Something in their description of the ailment triggered an association in Greatbatch's mind. He imagined the heart as a radio that was failing to transmit or receive a signal properly. He knew the history of modern electronics had been all about regulating the electrical signals passed between devices with ever more miraculous precision. Could you take all that knowledge and apply it to the human heart?

Greatbatch stored the idea in the back of his head for the next five years, where it lingered as a slow hunch. He moved to Buffalo, started teaching electrical engineering, and moonlighted at the Chronic Disease Institute. A physician at the institute recruited Greatbatch to help him engineer an oscillator that would record heartbeats using the new silicon transistors that were threatening to replace the vacuum tube. One day, while working on the device, Greatbatch happened to grab the wrong resistor. When he plugged it into the oscillator it began to pulse in a familiar rhythm. Thanks to Greatbatch's error, the device was simulating the beat of a human heart, not recording it. His mind flashed back to his conversation on the farm five years before. Here, at last, was the beginning of a device that could restore the faulty signal of an irregular heart, by shocking it back into sync at precise intervals. Within two years, Greatbatch and a Buffalo surgeon named William Chardack deployed the first implantable cardiac pacemaker on the heart of a dog. By 1960, the Greatbatch-Chardack pacemaker was pulsing steadily in the chests of ten human beings. Variations of Greatbatch's original design have now saved or prolonged millions of lives around the world.

Greatbatch's pacemaker is an instance where a great idea came—literally—from a novel combination of spare parts. Sometimes those novel combinations arrive courtesy of the random collisions of city streets or the dreaming brain. But sometimes they come from simple mistakes. You reach into the bag of resistors and pull out the wrong one, and four years later, you're saving someone's life. Yet error on its own is rarely enough. Greatbatch had his epiphany while hearing the reliable pulse of his oscillator because he'd been thinking about the irregular heartbeats as a signal transmission problem for five years. This, too, is a recurring pattern in the history of being wrong. The inventions of radiography, vulcanized rubber, and plastic all depended on generative mistakes that were generative precisely because they connected to slow hunches in the minds of their creators.

The British economist William Stanley Jevons, who had firsthand experience as an inventor himself, described the prominence of error in his *Principles of Science*, first published in 1874:

It would be an error to suppose that the great discoverer seizes at once upon the truth, or has any unerring method of divining it. In all probability the errors of the great mind exceed in number those of the less vigorous one. Fertility of imagination and abundance of guesses at truth are among the first requisites of discovery; but the erroneous guesses must be many times as numerous as those that prove well founded. The weakest analogies, the most whimsical notions, the most apparently absurd theories, may pass through the teeming brain, and no record remain of more than the hundredth part.

"The errors of the great mind exceed in number those of the less vigorous one." This is not merely statistics. It is not that the pioneering thinkers are simply more productive than less "vigorous" ones, generating more ideas overall, both good and bad. Some historical studies of patent records have in fact shown that overall productivity correlates with radical breakthroughs in science and technology, that sheer quantity ultimately leads to quality. But Jevons is making a more subtle case for the role of error in innovation, because error is not simply a phase you have to suffer through on the way to genius. Error often creates a path that leads you out of your comfortable assumptions. De Forest was wrong about the utility of gas as a detector, but he kept probing at the edges of that error, until he hit upon something that was genuinely useful. Being right keeps you in place. Being wrong forces you to explore.

Thomas Kuhn makes a comparable argument for the role of error in *The Structure of Scientific Revolutions*. Paradigm shifts, in Kuhn's argument, begin with anomalies in the data, when scientists find that their predictions keep turning out to be wrong. When Joseph Priestley first placed a mint plant in a bell jar to deprive it of oxygen, he expected that the plant would die, just as mice or spiders perished in the same circumstances. But he was wrong: the plant thrived. In fact, it thrived even if you burned all the oxygen out of the jar before placing the plant in it. Priestley's error energized him to investigate this strange behavior, and it ultimately led him to one of the founding discoveries of what we now call ecosystem science: the realization that plants expel oxygen as part of photosynthesis, and indeed have created much of the earth's atmosphere. As William James put it, "The error is needed to set off the truth, much as a dark background is required for exhibiting the brightness of a picture." When we're wrong, we have to challenge our assumptions, adopt new strategies. Being wrong on its own doesn't unlock new doors in the adjacent possible, but it does force us to look for them.

The trouble with error is that we have a natural tendency to dismiss it. When Kevin Dunbar analyzed the data from his *in vivo* studies of microbiology labs, one of his most remarkable findings was just how many experiments produced results that were genuinely unexpected. More than half of the data collected by the researchers deviated significantly from what they had predicted they would find. Dunbar found that the scientists tended to treat these surprising outcomes as the result of flaws in their experimental method: some kind of contamination of the original tissue perhaps, or a mechanical malfunction, or an error at the data-processing phase. They assumed the result was noise, not signal.

Transforming error into insight turned out to be one of the key functions of the lab conference. In Dunbar's research, outsiders working on different problems were much less likely to dismiss the apparent error as useless noise. Coming at the problem from a different perspective, with few preconceived ideas about what the "correct" result was supposed to be, allowed them to conceptualize scenarios where the mistake might actually be meaningful. As the science writer Jonah Lehrer has observed, this pattern appears in one of the great scientific breakthroughs of twentieth-century physics, the discovery of cosmic background radiation, which was mistaken for meaningless static by the astronomers Arno Penzias and Robert Wilson for more than a year, until a chance conversation with a Princeton nuclear physicist planted the idea that the noise was not the result of faulty equipment, but rather the still lingering reverberation of the Big Bang. Two brilliant scientists with great technological acumen stumble across evidence of the universe's origin—evidence that would ultimately lead to a Nobel Prize for both them—and yet their first reaction is: *Our telescope must be broken.*

About thirty years ago, a Berkeley psychology professor named Charlan Nemeth began investigating the relationship between noise, dissent, and creativity in group environments. One of Nemeth's early experiments assembled small groups of test subjects and showed them a series of slides, each of which was dominated by a single color. The subjects were asked to evaluate the color and the brightness of each slide. After they had analyzed the slides, Nemeth asked them to free-associate on the color they had perceived in the slides.

There are few actions as commonly connected to the pursuit of creativity as free-associating. Trying to come up with a new slogan for a detergent? Struggling for a new perspective on your memories of childhood trauma? Compiling thoughts for a sonnet? Free-associating, we are told, will help us find our answer.

But psychologists have long been in on the joke that humans free-associate in absurdly predictable ways. Take a hundred Americans off the street and ask them to free-associate on the word "green" and forty of them will say "grass." Another forty will offer up another color—"red" or "yellow" or "blue"—or the word "color" itself. The more creative associations only emerge when you get to the bottom 20 percent of responses, the long tail of free association, where words like "Ireland," or "money," or "leaves" appear. Ask them to free-associate on the word "blue" and you'll see the same pattern: 80 percent will suggest either another color or the word "sky," and the last 20 percent of associations will be scattered across dozens of less predictable responses: "jeans," "lake," or "lonely."

Psychologists have assembled immense probability tables that document the patterns of free association for hundreds of words. These norms of association give them a stable yardstick for measuring creative thinking in different environments. Some situations cause people to get even more predictable with their associations, offering up "grass" and "blue" like obedient robots. But other situations can push their associations down the tail of the distribution, into the more eclectic zone of "Ireland" and "money." Individuals who are unusually creative tend to generate more original associations when tested.

Charlan Nemeth's experiment was a perfect embodiment of this predictability. Blue slides triggered utterly conventional word asso-

ciations: "sky" and "green" and "color" dominated, while the more
innovative associations were restricted to the bottom 20 percent.

But then Nemeth ran another version of the experiment, this
time with a twist. She showed the same slides to small groups of
subjects—only, in this version, she secretly seeded each group with
a handful of actors who were instructed to describe each slide inac-
curately, as if it were a different color. The real test subjects correctly described the blue slides as blue and were surprised to find
that their peers somehow saw the very same color and perceived it
as green.

When Nemeth took this cohort (that is, the test subjects minus
the actors) and asked them to free-associate on the color names they
had mentioned, the words they came up with were markedly different from the earlier group's. Some of them dutifully suggested "sky,"
like normal respondents, but the sort of associations that usually
resided in the creative tail of the distribution—"jazz" or "jeans"—
were far more numerous. In other words, when subjects were exposed to *inaccurate* descriptions of the slides, they became more
creative. Associations that traditionally lay on the fringes of the probability table suddenly became mainstream. Nemeth had deliberately
introduced noise into the decision-making process, and what she
found ran directly counter to our intuitive assumptions about truth
and error. The groups that had been deliberately contaminated with
erroneous information ended up making more original connections
than the groups that had only been given pure information. The
"dissenting" actors prodded the other subjects into exploring new
rooms in the adjacent possible, even though they were, technically
speaking, adding incorrect data to the environment.

Nemeth has gone on to document the same phenomenon at work in dozens of different environments: mock juries, boardrooms, academic seminars. Her research suggests a paradoxical truth about innovation: good ideas are more likely to emerge in environments that contain a certain amount of noise and error. You would think that innovation would be more strongly correlated with the values of accuracy, clarity, and focus. A good idea has to be *correct* on some basic level, and we value good ideas because they tend to have a high signal-to-noise ratio. But that doesn't mean you want to cultivate those ideas in noise-free environments, because noise-free environments end up being too sterile and predictable in their output. The best innovation labs are always a little contaminated.

The next time you visit a zoo or a natural history museum and survey the extraordinary diversity of the organisms on our planet, pause for a second to remind yourself that all this variation—the elephant tusks and peacock tails and human neocortices—was made possible, in part, by error. Without noise, evolution would stagnate, an endless series of perfect copies, incapable of change. But because DNA is susceptible to error—whether mutations in the code itself or transcription mistakes during replication—natural selection has a constant source of new possibilities to test. Most of the time, these errors lead to disastrous outcomes, or have no effect whatsoever. But every now and then, a mutation opens up a new wing of the adjacent possible. From an evolutionary perspective, it's not enough to say "to err is human." Error is what made humans possible in the first place.

The prominence of random mutation in our evolutionary history has long been associated with Darwin's original theory, but the truth is that Darwin himself had a hard time accepting the premise that undirected random variation could produce the marvelous innovations of life. When Darwin first outlined the theory of natural selection as the "preservation of favourable variations and rejection of injurious variations" in *On the Origin of Species*, he lacked a convincing theory about where all those variations came from. In *Origin*, he generally writes about them as though they are random, in part because he is explicitly trying to shed the Lamarckian notion of directed variation, where new innovations—the giraffe's long neck being the canonical example—are generated by activity during the organism's lifetime, and then passed down to the next generation. But during the decade that followed, Darwin retreated from the cliff of random variation and developed a theory called pangenesis, first published in his 1868 book, *The Variation of Animals and Plants under Domestication*. Pangenesis dialed back the noise of Darwin's original theory, introducing a complex mechanism for heredity that created a kind of directed variation. In Darwin's theory, each cell in the body released hereditary particles called gemmules that collected in the germ cells of the organism. A particular organ or limb that was heavily used during the lifetime of the animal would release more gemmules, and thus shape the physiology of the next generation. Pangenesis was well received at the time Darwin proposed it, but the modern science of genetics would ultimately reveal it to be entirely false. It would prove to be the most egregious mistake of his scientific career. In a sense, Darwin's greatest error was his failure to understand the protean force of error.

Too much error is deadly, of course, which is why your cells contain elaborate mechanisms for repairing damaged DNA and for ensuring that the transcoding process is accurate down to the last nucleotide. An organism that constantly rescrambled the genetic code passed down to its descendants would be more innovative in its offspring, but only in the sense that those offspring would find many novel ways to perish before or shortly after birth. No parents want genetic mutations in their child. But as a species we have been dependent on mutation.

That dependence is why some scientists have argued that natural selection has gravitated toward a small but stable error rate in DNA transcoding, that evolution has, in a sense, "tuned" the error rate to the optimal balance between too much mutation and too much stability. One might think, given the severe threat associated with transcoding errors, that there would be extraordinary selection pressure to make the DNA repair system foolproof. Parents who made perfect copies of their germ cells would have healthier offspring, while parents with faulty DNA repair would have fewer surviving offspring, thanks to their higher mutation rates. Over time, the genes for foolproof DNA repair would spread through the society at large. The complexity of the DNA repair system suggests that evolution did largely follow this path, only it stopped short of eliminating error altogether. Our cells appear to be designed to leave the door for mutation ever so slightly open, just enough to let a small trickle of change and variation in, without catastrophic effects for the population as a whole. Recent studies suggest that the mutation rate in human germ cells is roughly one in thirty million base pairs, which means each time parents pass their DNA on to a child, that genetic inheritance comes with roughly 150 mutations. Much of the machinery in our cells is devoted to preserving and reproducing the signal of the genetic code. But evolution has still made room for noise.

Is that error rate the result of selection pressures, or just a reflection of the fact that evolution is not perfect? Humans have relatively good vision, as mammals go, but we can't read magazine text from five hundred feet. That's not necessarily a sign that there is something adaptive about that limitation; it's more likely that it's hard to engineer an eye that can see that well; and, as powerful as evolution is, it can't do everything. Presumably we would have been more evolutionarily "fit" if we'd been able to run a hundred miles per hour, too, but the restrictions of our bone and muscle structure kept us from being able to outrace the cheetahs. Why couldn't the same be true of our imperfect DNA repair system?

It may well be that perfect replication is simply an ideal limit that natural selection can only approach asymptotically. For our purposes, it doesn't really matter whether selection has actively tuned our DNA repair systems for a certain level of noise or whether they simply fell short of their "goal" of perfect reproduction. One way or another, the noise had to be preserved, because without it, evolution would grind to a halt. But the tuning hypothesis has had some tantalizing research on its side of late. Bacteria have much higher mutation rates than multicellular life-forms, which suggests that the tolerance for error varies according to the specific circumstances of different organisms. One study by Susan Rosenberg at Baylor College found that bacteria increased their mutation rates dramatically when confronted with the "stress" of low energy supplies. When the living is good, Rosenberg's research suggests, bacteria have less of a need for high mutation rates, because their current strategies are well adapted to their environment. But when the environment grows more hostile, the pressure to innovate—to stumble across some new way of eking out a living in a resource-poor setting—shifts the balance of risk versus reward involved in mutation. The risk of your offspring dying from some deadly mutation doesn't look quite as bad if they're going to die of starvation anyway. And if one of those mutations helps the bacteria use the limited energy resources more efficiently, the new gene will quickly spread through the population as the nonmutated bacteria die off.

In a sense, Rosenberg's mutating bacteria are following a strategy similar to what the water fleas adopted in their oscillation between sexual and asexual reproduction. When the going gets tough, life tends to gravitate toward more innovative reproductive strategies, sometimes by introducing more noise into the signal of genetic code, and sometimes by allowing genes to circulate more quickly through the population.

Sex and error turn out to have a long interconnected history, which is probably not news to those who remember their college love life. One of the key advantages to sexual reproduction is that it enables mutated genes to break off from the genes that produce higher rates of mutation. Picture a bacterium that possesses a gene that inhibits its DNA repair slightly, increasing its overall mutation rate. Most of those mutations will be inconsequential or downright lethal, but imagine one day it hits the jackpot and stumbles across a mutation that *increases* its reproductive fitness—say, for the sake of argument, that it enables the organism to detect food sources more efficiently. Our fortunate bacterium splits in two and passes its genes on to the next generation. The trouble is, that next generation gets a mixed inheritance: it inherits the new scavenging gene, but it also inherits the gene that produces higher mutation

rates. Because negative mutations are much more likely than positive ones, over generations the advantages of the scavenging gene
get overwhelmed by the noise introduced by the gene that causes
higher mutation rates. But if our lucky bacterium could suddenly
switch to sexual reproduction, as the water flea does, the outcome
might be very different, because in sexual reproduction you only
pass along half your genes to your offspring. The next generation
can inherit her father's knack for scavenging and her mother's gift
for accurate DNA repair.

We have already explored some of the reasons why evolution would gravitate toward the far more complicated system of
sexual reproduction: it allows potentially useful innovations to
spread through the population and occasionally collide and join
forces with other innovations. But when you think about sex in the
context of those mutation and scavenging genes, it becomes clear
that so much of life on earth embraced sexual reproduction for
another reason: because sex helped harness the generative power of
error while mitigating the risks. Sex keeps the door to the adjacent
possible open by just a crack, so that we can adapt to the changing
pressures or opportunities of our environment. By keeping the
opening so narrow, it also keeps mutation rates in check, which is
one crucial reason the asexual bacteria have such markedly higher
error rates than the multicellular life. Sex lets us learn from the
mistakes of our genes.

It's this complicated relationship between accuracy and error,
between signal and noise, that explains Charlan Nemeth's research
on free association and jury deliberation. When one of our peers
calls the blue painting green, or comes to the defense of a suspect
who is clearly guilty, he or she is, technically speaking, introducing

more inaccurate information to the environment. But that noise makes the rest of us smarter, more innovative, precisely because we're forced to rethink our biases, to contemplate an alternate model in which the blue paintings are, in fact, green. Being correct is like the phase-lock states of the human brain, all the neurons firing in perfect synchrony. We need the phase-lock state for the same reason we need truth: a world of complete error and chaos would be unmanageable, on a social and a neurochemical level. (Not to mention genetic.) But leaving some room for generative error is important, too. Innovative environments thrive on useful mistakes, and suffer when the demands of quality control overwhelm them. Big organizations like to follow perfectionist regimes like Six Sigma and Total Quality Management, entire systems devoted to eliminating error from the conference room or the assembly line, but it's no accident that one of the mantras of the Web startup world is *fail faster*. It's not that mistakes are the goal—they're still mistakes, after all, which is why you want to get through them quickly. But those mistakes are an inevitable step on the path to true innovation. Benjamin Franklin, who knew a few things about innovation himself, said it best: "Perhaps the history of the errors of mankind, all things considered, is more valuable and interesting than that of their discoveries. Truth is uniform and narrow; it constantly exists, and does not seem to require so much an active energy, as a passive aptitude of soul in order to encounter it. But error is endlessly diversified."

VI.

EXAPTATION

Two years before Pliny the Elder died, during a daring rescue of friends after the eruption of Mount Vesuvius, the legendary Roman historian and scholar completed his proto-encyclopedia, *Naturalis Historiae.* In it he tells the story of a device winemakers had recently invented, a new kind of press that employed a screw to "concentrate pressure upon broad planks placed over the grapes, which are covered also with heavy weights above." There is some debate among scholars over whether Pliny may have been rooting for the home team in attributing the invention to his compatriots, since evidence for the use of screw presses in producing wines and olive oils dates back several centuries, to the Greeks. But whatever the exact date of its origin, the practical utility of the screw press, unlike so many great ideas from the Greco-Roman period, ensured that it survived intact through the Dark Ages. When the Renaissance finally blossomed, more than a millennium after Pliny's de-

mise, Europe had to rediscover Ptolemaic astronomy and the secrets of building aqueducts. But they didn't have to relearn how to
press grapes. In fact, they'd been tinkering steadily with the screw
press all along, improving on the model, and optimizing it for the
mass production of wines. By the mid-1400s, the Rhineland region of Germany, which historically had been hostile to viticulture
for climate reasons, was now festooned with vine trellises. Fueled
by the increased efficiency of the screw press, German vineyards
reached their peak in 1500, covering roughly four times as much
land as they do in their current incarnation. It was hard work producing drinkable wine in a region that far north, but the mechanical efficiency of the screw press made it financially irresistible.

Sometime around the year 1440, a young Rhineland entrepreneur began tinkering with the design of the wine press. He was
fresh from a disastrous business venture manufacturing small mirrors with supposedly magical healing powers, which he intended to
sell to religious pilgrims. (The scheme got derailed, in part by bu-
bonic plague, which dramatically curtailed the number of pilgrims.) The failure of the trinket business proved fortuitous,
however, as it sent the entrepreneur down a much more ambitious
path. He had immersed himself in the technology of Rhineland
vintners, but Johannes Gutenberg was not interested in wine. He
was interested in words.

As many scholars have noted, Gutenberg's printing press was
a classic combinatorial innovation, more bricolage than breakthrough. Each of the key elements that made it such a transformative machine—the movable type, the ink, the paper, and the
press itself—had been developed separately well before Gutenberg
printed his first Bible. Movable type, for instance, had been inde-

pendently conceived by a Chinese blacksmith named Pi Sheng four centuries before. But the Chinese (and, subsequently, the Koreans) failed to adapt the technology for the mass production of texts, in large part because they imprinted the letterforms on the page by hand rubbing, which made the process only slightly more efficient than your average medieval scribe. Thanks to his training as a goldsmith, Gutenberg made some brilliant modifications to the metallurgy behind the movable type system, but without the press itself, his meticulous lead fonts would have been useless for creating mass-produced Bibles.

An important part of Gutenberg's genius, then, lay not in conceiving an entirely new technology from scratch, but instead from *borrowing* a mature technology from an entirely different field, and putting it to work to solve an unrelated problem. We don't know exactly what chain of events led Gutenberg to make that associative link; few documentary records remain of Gutenberg's life between 1440 and 1448, the period during which he assembled the primary components of his invention. But it is clear that Gutenberg had no formal experience pressing grapes. His radical breakthrough relied, instead, on the ubiquity of the screw press in Rhineland winemaking culture, and on his ability to reach out beyond his specific field of expertise and concoct new uses for an older technology. He took a machine designed to get people drunk and turned it into an engine for mass communication.

Evolutionary biologists have a word for this kind of borrowing, first proposed in an influential 1971 essay by Stephen Jay Gould and Elisabeth Vrba: *exaptation*. An organism develops a trait optimized for a specific use, but then the trait gets hijacked for a completely different function. The classic example, featured prominently in Gould and Vrba's essay, is bird feathers, which we believe initially evolved for temperature regulation, helping nonflying dinosaurs from the Cretaceous period insulate themselves against cold weather. But when some of their descendants, including a creature we now call *Archaeopteryx*, began experimenting with flight, feathers turned out to be useful for controlling the airflow over the surface of the wing, allowing those first birds to glide.

The initial transformation is almost accidental: a tool sculpted by evolutionary pressures for one purpose turns out to have an unexpected property that helps the organism survive in a new way. But once that new property gets put to use, once *Archaeopteryx* starts using its feathers to glide, the trait evolves according to a different set of criteria. All flight feathers, for instance, have pronounced asymmetry to them: the vane on one side of the central shaft is larger than the vane on the opposite side. This lets the feather act as a kind of airfoil, providing lift during the flapping of wings. Birds that fly at unusually high velocities, like hawks, have more extreme asymmetries than slower birds. Yet down feathers that simply provide insulation are perfectly symmetrical. When your feathers are there just to keep you warm, there's no advantage to building slightly off-kilter feathers. Mutations or other general variability in the gene pool inevitably produces feathers that are slightly less symmetrical than average, but those traits don't intensify and spread across generations because they don't convey any reproductive advantage over normal feathers. But once flight speed becomes a property with major implications for survival, those asymmetrical vanes turn out to be extremely useful. Where asymmetry had previously drifted in and out of the gene pool, natural selection now begins sculpting those feathers to make them more aerodynamic. A feather *adapted* for warmth is now *exapted* for flight.

The concept of exaptation is crucial in rebutting the classic biblical argument (now often termed "intelligent design") against Darwinism, one that dates back to the furor surrounding the publication of *On the Origin of Species* itself: if extraordinary examples of natural engineering like eyes or wings are *not* the product of an intelligent creator, then how could these traits have survived through what must have been a pronounced developmental state of nonfunctionality? As the wing evolves, by definition it has to go through a long period where it's utterly useless at flying. (As the saying goes: "What good is 5 percent of a wing?") Because natural selection doesn't "know" that it's trying to build a wing, it can't push those emerging wings toward the ultimate goal of flying the way a mechanical engineer can continue tinkering with a toy airplane until it successfully takes to the air. If your aspiring wing doesn't help you to fly, and thus outmaneuver your predators or discover new sources of food, the new mutations that made that appendage slightly more winglike won't be more likely to spread through the population. Natural selection doesn't give good grades for effort.

But when you think about evolutionary innovation in terms of exaptations, the story becomes far less mysterious. Once again, chance and happy accidents are central to the narrative: random mutations lead to the evolution of feathers selected for warmth, and by chance those feathers turn out to be useful for flying, particularly after they've been modified to create an airfoil. Sometimes those exaptations become possible because other exaptations are happening within the species: the wing itself is thought to be an exaptation of a dinosaur wrist bone originally adapted for greater flexibility. When Gould offered up his tires-to-sandals metaphor, he was essentially talking about the way in which exaptations have defined the paths of evolutionary innovation: new abilities and traits come about not because there is some inexorable march toward more and more complexity in the biosphere, but rather because natural selection has the Nairobi cobbler's instinct for taking old parts and putting them to new uses.

Oftentimes those new uses become possible thanks to external changes in an organism's environment. When the lobe-finned fish Sarcopterygii first began exploring life at the water's edge, 400 million years ago, the creature had a small swim fan at the end of its fin, supported by narrow rays of bone. As its descendants began to spend more time away from the water, exploiting the copious energy sources of the plants and arthropods that had already conquered life onland, the tip of the lobe-fin turned out to be useful for an activity that acquatic life had rendered unthinkable: walking. Before long, natural selection had refashioned the swim fan into an autopod, the basic architecture of all mammalian ankles and feet. Over time, the autopod itself would be exapted in numerous ways: creating primate hands and fingers optimized for grasping, or those *Archaeopteryx* wings. In some cases, the autopod was even exapted back to its ancient swim-fan origins, as in the flippers of seals and sea lions.

If mutation and error and serendipity unlock new doors in the biosphere's adjacent possible, exaptations help us explore the new possibilities that lurk behind those doors. A match you light to illuminate a darkened room turns out to have a completely different use when you open a doorway and discover a room with a pile of logs and a fireplace in it. A tool that helps you see in one context ends up helping you keep warm in another. That's the essence of exaptation.

It's tempting to assume that the machinery of cultural innovation is closer to that engineer tinkering with her model airplane than it is to the lucky *Archaeopteryx* leaping off the treetop and discovering that its feathers are more than just a down jacket. No one contests the role of intelligent design in the history of human culture. But the history of human creativity abounds with exaptations. In the early 1800s, a French weaver named Joseph-Marie Jacquard developed the first punch cards to weave complex silk patterns with mechanical looms. Several decades later, Charles Babbage borrowed Jacquard's invention to program the Analytical Engine. Punch cards would remain crucial to programmable computers until the 1970s. Lee de Forest created the Audion with one clear aim: to create a device that would detect electromagnetic signals and amplify them. It never occurred to him that the triode architecture could just as easily be applied to the problem of building a hydrogen bomb. In evolutionary terms, the vacuum tube was originally adapted to make signals louder, but it was eventually exapted to turn those signals into information: zeros and ones that could be manipulated in astonishing ways. A Fender guitar amp from the fifties that relied on a vacuum tube to boost the signal of the first rock-and-roll guitarists was, ultimately, a variation on de Forest's original amplification theme. But those 17,000 vacuum tubes inside ENIAC, doing the math on the physics of a hydrogen bomb—they were serving a function that never crossed de Forest's mind, however imaginative it might have been. Today, emerging patent marketplaces like Nike's GreenXchange are enabling commercial exaptations that would have been unthinkable in the fortified environment of traditional R&D labs.

The history of the World Wide Web is, in a sense, a story of continuous exaptation. Tim Berners-Lee designs the original protocols with a specifically academic environment in mind, creating a platform for sharing research in a hypertext format. But when the first Web pages crawl out of that scholarly primordial soup and begin to engage with ordinary consumers, Berners-Lee's invention turns out to possess a remarkable number of unanticipated qualities. A platform adapted for scholarship was exapted for shopping, and sharing photos, and watching pornography—along with a thousand other uses that would have astounded Berners-Lee when he created his first HTML-based directories in the early nineties. When Sergey Brin and Larry Page decided to use links between Web pages as digital votes endorsing the content of those pages, they were exapting Berners-Lee's original design: they took a trait adapted for navigation—the hypertext link—and used it as a vehicle for assessing quality. The result was PageRank, the original algorithm that made Google into the behemoth that it is today.

The literary historian Franco Moretti has persuasively documented the role of exaptation in the evolution of the novel. An author conceives a new kind of narrative device to address a specific, local need in a work he or she is writing. Something about the device resonates with other authors, and it begins to circulate through the literary gene pool. And then, as the literary environment changes and new imaginative possibilities become necessary, the device turns out to have a different function, far removed from its original use. The French novelist Edouard Dujardin first uses the "stream of consciousness" technique in his 1888 novel *Les Lauriers sont coupés*; in Dujardin's rendition, the technique is restricted to short periods of introspection between the main events of the story, brief parentheses within the plot. But three decades later, James Joyce would take the device and transform it into the most memorable and mesmerizing perceptual modes, using the device in his novel *Ulysses* to capture the churn and distractibility of mental life in a bustling city. When Dickens conjured up his Inspector Bucket to weave together the multiplying strands of metropolitan coincidence in *Bleak House*, he had no idea his contrivance would help create a whole new genre of detective fiction, one that would extend all the way from Wilkie Collins's *The Moonstone* to Sherlock Holmes to *Murder, She Wrote*. New genres need old devices.

Rhetorical or figurative exaptations are not the exclusive property of the arts. The history of scientific and technological innovation abounds with them as well. In *The Act of Creation*, Arthur Koestler argued that "all decisive events in the history of scientific thought can be described in terms of mental cross-fertilization between different disciplines." Concepts from one domain migrate to another as a kind of structuring metaphor, thereby unlocking some secret door that had long been hidden from view. In his memoirs, Francis Crick reports that he first hit upon the complementary replication system of DNA—each base A matched with a T, and each C with a G—by thinking of the way a work of sculpture can be reproduced by making an impression in plaster, and then using that impression, when dry, as a mold to create copies. Johannes Kepler credited his laws of planetary motion to a generative metaphor imported from religion; he imagined the sun, stars, and the dark space between them as the celestial equivalents of the Father, Son, and

Holy Ghost. When computer science pioneers like Doug Engelbart and Alan Kay invented the graphical interface, they imported a metaphor from the real-world environment of offices: instead of organizing information on the screen as a series of command-line inputs, the way a programmer would, they borrowed the iconography of a desktop with pieces of paper stacked on it. Kekulé didn't think the benzene molecule was literally a snake from Greek mythology, but his knowledge of that ancient symbol helped him solve one of the essential problems of organic chemistry.

In the early 1970s, a Berkeley sociologist named Claude Fischer began investigating the social effects of living in dense urban centers. The topic was one that had long interested urban theorists, most famously in Louis Wirth's classic essay from 1938, "Urbanism as a Way of Life," which argued that metropolitan living led toward social disorganization and alienation, the social ties and comforts of smaller communities breaking down in the tumult of the big city. Wirth's argument had not aged well—it turned out that densely populated neighborhoods had very complex and rich social bonds if one looked for them—and so Fischer set out to determine what social patterns were truly precipitated by the environment of large cities. His research led him to one overwhelming conclusion, published in a seminal paper in 1975: big cities nurture *subcultures* much more effectively than suburbs or small towns.

Lifestyles or interests that deviate from the mainstream need critical mass to survive; they atrophy in smaller communities not because those communities are more repressive, but rather because the odds of finding like-minded people are much lower with a smaller pool of individuals. If one-tenth of one percent of the population are passionately interested in, say, beetle collecting or improv theater, there might only be a dozen such individuals in a midsized town. But in a big city there might be thousands. As Fischer noted, that clustering creates a positive feedback loop, as the more unconventional residents of the suburbs or rural areas migrate to the city in search of fellow travelers. "The theory . . . explains the 'evil' and 'good' of cities simultaneously," Fischer wrote. "Criminal unconventionality and innovative (e.g., artistic) unconventionality are both nourished by vibrant subcultures." Poetry collectives and street gangs might seem miles apart on the surface, but they each depend on the city's capacity for nurturing subcultures.

The same pattern holds true for trades and businesses in large cities. As Jane Jacobs observed in *The Death and Life of Great American Cities*: "The larger a city, the greater the variety of its manufacturing, and also the greater both the number and the proportion of its small manufacturers."

> Towns and suburbs, for instance, are natural homes for huge supermarkets and for little else in the way of groceries, for standard movie houses or drive-ins and for little else in the way of theater. There are simply not enough people to support further variety, although there may be people (too few of them) who would draw upon it were it there. Cities, however, are the natural homes of supermarkets and standard movie houses plus delicatessens, Viennese bakeries, foreign groceries, art movies, and so on, all of which can be found co-existing, the standard with the strange, the large with the small. Wherever lively and popular parts of cities are found, the small much outnumber the large.

Both Fischer and Jacobs emphasize the fertile interactions that occur between subcultures in a dense city center, the inevitable spillover that happens whenever human beings crowd together in large groups. Subcultures and eclectic businesses generate ideas, interests, and skills that inevitably diffuse through the society, influencing other groups. As Fischer puts it, "The larger the town, the more likely it is to contain, in meaningful numbers and unity, drug addicts, radicals, intellectuals, 'swingers,' health-food faddists, or whatever; and the more likely they are to influence (as well as offend) the conventional center of the society."

Cities, then, are environments that are ripe for exaptation, because they cultivate specialized skills and interests, and they create a liquid network where information can leak out of those subcultures, and influence their neighbors in surprising ways. This is one explanation for superlinear scaling in urban creativity. The cultural diversity those subcultures create is valuable not just because it makes urban life less boring. The value also lies in the unlikely migrations that happen between the different clusters. A world where a diverse mix of distinct professions and passions overlap is a world where exaptations thrive.

Those shared environments often take the form of a real-world public space, what the sociologist Ray Oldenburg famously called the "third place," a connective environment distinct from the more insular world of home or office. The eighteenth-century English coffeehouse fertilized countless Enlightenment-era innovations; everything from the science of electricity, to the insurance industry, to democracy itself. Freud maintained a celebrated salon Wednesday nights at 19 Berggasse in Vienna, where physicians, philosophers, and scientists gathered to help shape the emerging field of psychoanalysis. Think, too, of the Paris cafés where so much of modernism was born; or the legendary Homebrew Computer Club in the 1970s, where a ragtag assemblage of amateur hobbyists, teenagers, digital entrepreneurs, and academic scientists managed to spark the personal computer revolution. Participants flock to these spaces partly for the camaraderie of others who share their passions, and no doubt that support network increases the engagement and productivity of the group. But encouragement does not necessarily lead to creativity. *Collisions* do—the collisions that happen when different fields of expertise converge in some shared physical or intellectual space. That's where the true sparks fly. The modernism of the 1920s exhibited so much cultural innovation in such a short period of time because the writers, poets, artists, and architects were all rubbing shoulders at the same cafés. They weren't off on separate islands, teaching creative writing seminars or doing design reviews. That physical proximity made the space rich with exaptation: the literary stream of consciousness influencing the dizzying new perspectives of cubism; the futurist embrace of technological speed in poetry shaping new patterns of urban planning.

Exaptation also prospers on another scale: the shared *media* environment of a physical community. In the late 1970s, the British musician and artist Brian Eno moved to New York City for the first time. He took over a flat in a converted town house in the heart of the Village. The city was at the height—or more like the nadir—of its rioting, Son of Sam–fearing, bankruptcy-flirting madness. Still, having spent time in 1970s London and Berlin, Eno was well acclimated to urban anarchy. In fact, the most jarring contrast to his European past was the turbulent mix of voices on the radio. After years of listening to the somber, professional voices of the BBC, the outlandish rants of American radio seemed to Eno like a new universe of insanity.

And so he started taping them. Like many experimental musicians at that point, Eno had been exploring the possibilities of using tape loops as a musical instrument. ("The tape recorder was always the instrument I felt most comfortable with," he once said in an interview. "Keyboards after that, with bass as a distant third.") The Beatles had reserved the longest track on the White Album for Lennon's tape-loop collage "Revolution #9," and the proto-synthesizer Mellotron, developed in the mid-sixties, had separate tape loops set up to be triggered by individual keys on the keyboard. But none of those experiments had ever really employed the spoken voice as a harmonic or percussive element. The drones and murmurs of "Revolution #9" were, after all, barely musical by traditional standards. But Eno's hours with the evangelists and the anarchists and the shock-jocks-in-embryo had lodged those voices in his head, and as he began work on a collaboration with David Byrne, he started to toy with the idea of exploring their musical possibilities. The result was *My Life in the Bush of Ghosts*, an utterly original mix of African rhythm sections and oddball acoustic instruments, but notably missing Byrne's taut New Wave vocal stylings—so prominently featured in the Talking Heads albums the two had previously collaborated on. Instead of traditional singing, Byrne and Eno built the songs around the layered, looped ensemble of spoken words that Eno had grabbed from the airwaves. It was a case study in creative exaptation: words designed in one medium to spread the word of Jesus, or to thunder against the military-industrial complex, migrated over into a new environment and became, against all odds, music.

My Life in the Bush of Ghosts marked the birth of a certain historically crucial kind of musical borrowing: it was not just a new music, but a whole new way of thinking about what music could be built out of. (Not unlike the way Marcel Duchamp and his fellow surrealists had changed our understanding of what art could be made of fifty years before.) Several years later, when Public Enemy producer Hank Shocklee sat down to record the album *It Takes a Nation of Millions to Hold Us Back*, he deliberately mimicked the layered, percussive vocal samples of Eno and Byrne's production. *It Takes a Nation* went on to become one of the most sonically influential records of its decade, reverberating through the wider culture—in everything from cell phone jingles to billboard chart-toppers to avant-garde experimentation—just as *Highway 61 Revisited* and *Pet Sounds* had done a generation before. Eno's original innovation was brilliant, to be sure, and from a distance it almost looks like the classic "lone genius" eureka moment: the innovator locked away in his lab, stumbling across an idea that would transform the wider culture. But it is crucial to the story that Eno was not, technically speaking, alone with his tape recorder: he was tapped into a network of wildly different voices, all of them ranting at different frequencies. Eno didn't need a coffeehouse. He had AM radio.

In the late nineties, a Stanford Business School professor named Martin Ruef decided to investigate the relationship between business innovation and diversity. Ruef was interested in the coffeehouse model of diversity, not the "melting pot" political kind:

the diversity of professions and disciplines, not of race or sexual orientation. Ruef interviewed 766 graduates of the school who had gone on to have entrepreneurial careers. He created an elaborate system for scoring innovation based on a combination of factors: the introduction of new products, say, or the filing of trademarks and patents. And then he tracked each graduate's social network—not just the number of acquaintances but the *kind* of acquaintances they had. Some graduates had large social networks that were clustered within their organization; others had small insular groups dominated by friends and family. Some had wide-ranging connections with acquaintances outside their inner circle of friends and colleagues.

What Ruef discovered was a ringing endorsement of the coffeehouse model of social networking: the most creative individuals in Ruef's survey consistently had broad social networks that extended outside their organization and involved people from diverse fields of expertise. Diverse, horizontal social networks, in Ruef's analysis, were *three times* more innovative than uniform, vertical networks. In groups united by shared values and long-term familiarity, conformity and convention tended to dampen any potential creative sparks. The limited reach of the network meant that interesting concepts from the outside rarely entered the entrepreneur's consciousness. But the entrepreneurs who built bridges outside their "islands," as Ruef called them, were able to borrow or co-opt new ideas from these external environments and put them to use in a new context. A similar study, conducted by a University of Chicago business school professor named Ronald Burt, looked at the origin of good ideas inside the organizational network of the Raytheon Corporation. Burt found that innovative thinking was much more likely to emerge from individuals who bridged "structural holes" between tightly knit clusters. Employees who primarily shared information with people in their own division had a harder time coming up with useful suggestions for Raytheon's business, when measured against employees who maintained active links to a more diverse group.

To a certain extent, Ruef's and Burt's research is a validation of the celebrated "strength of weak ties" argument first proposed by Mark Granovetter, and popularized by Malcolm Gladwell in *The Tipping Point*. But looking at the weak ties of an extended social network through the lens of exaptation changes the picture in an important way: it is not merely that weak ties allow information to travel throughout a network more efficiently—that is, without becoming trapped on the remote island of a close-knit group. From the perspective of innovation, it's even more important that the information arriving from one of those weak ties is coming from a different context, what the innovation scholar Richard Ogle calls an "idea-space": a complex of tools, beliefs, metaphors, and objects of study. A new technology developed in one idea-space can migrate over to another idea-space through these long-distance connections; in that new environment, the technology may turn out to have unanticipated properties, or may trigger a connection that leads to a new breakthrough. The value of the weak tie lies not just in the speed with which it transmits information across a network; it also promotes the exaptation of those ideas. Gutenberg was trained as a metallurgist, but he had weak ties to the vintners of Rhineland Germany. Without that link, he would have been merely a pioneering typesetter, making an incremental improvement on Pi Sheng's movable type. By not restricting himself exclusively to the island of metallurgy, he became something much more important: a printer.

The model of weak-tie exaptation also helps us understand the classic story of twentieth-century scientific epiphany: Watson and Crick's discovery of the double-helix structure of DNA. As Ogle and others have noted, in the small scientific community working on the problem of DNA in the early 1950s, the person who had the clearest and most direct view of the molecule itself was neither James Watson nor Francis Crick. It was, instead, a biophysicist at London University named Rosalind Franklin, who was using state-of-the-art X-ray crystallography to study the mysterious strands of DNA. But Franklin's vision was limited by two factors. First, there was the imperfect state of the X-ray technology, which only gave her hints about the helix structure and base-pair symmetry. But Franklin was also limited by the conceptual island on which she based her work. Her approach was purely inductive: master the X-ray technology and then use the information collected to build a model of DNA. ("We're going to let the data tell us the structure," she famously told Crick.) But to "see" the double-helix in the early 1950s took something more than just analyzing it in an X-ray machine. To solve the mystery, Watson and Crick had to piece it together with tools drawn from multiple disciplines: biochemistry, genetics, information theory, and mathematics, not to mention Franklin's X-ray images. Even Crick's sculpture metaphor proved crucial to cracking the code. Next to Franklin, Watson and Crick seemed almost dilettantes and dabblers: Crick had switched from physics to biology in his graduate years; neither had a comprehensive grasp of biochemistry. But DNA was not a problem that could be solved within a single discipline. Watson and Crick had to borrow from other domains to make sense of the molecule. As Ogle puts it, "Once key ideas from idea-spaces that otherwise had little contact with one another were connected, they began, quasi-autonomously, to make new sense in terms of one another, leading to the emergence of a whole that was more than the sum of its parts." It is a fitting footnote to the story that Watson and Crick were notorious for taking long, rambling coffee breaks, where they tossed around ideas in a more playful setting outside the lab—a practice that was generally scorned by their more fastidious colleagues. With their weak-tie connections to disparate fields, and their exaptative intelligence, Watson and Crick worked their way to a Nobel Prize in their own private coffeehouse.

The coffeehouse model of creativity helps explain one of those strange paradoxes of twenty-first-century business innovation. Even as much of the high-tech culture has embraced decentralized, liquid networks in their approach to innovation, the company that is consistently ranked as the most innovative in the world—Apple— remains defiantly top-down and almost comically secretive in its development of new products. You won't ever see Steve Jobs or Jonathan Ive crowdsourcing development of the next-generation iPhone. If open and dense networks lead to more innovation, how can we explain Apple, which on the spectrum of openness is far closer to Willy Wonka's factory than it is to Wikipedia? The easy answer is that Jobs and Ive simply possess a collaborative genius that has enabled the company to ship such a reliable stream of revolutionary products. No doubt both men are immensely talented at what they do, but neither of them can design, build, program, and market a product as complex as the iPhone on their own, the way Jobs and Steve Wozniak crafted the original Apple personal computer in the now-legendary garage. Apple clearly has unparalleled leadership, but there must also be something in the environment at Apple that is allowing such revolutionary ideas to make it to the marketplace.

As it turns out, while Apple has largely adopted a fortress mentality toward the outside world, the company's internal development process is explicitly structured to facilitate clash and connection between different perspectives. Jobs himself has taken to describing their method via the allegory of the concept car. You go to an auto show and see some glamorous and wildly innovative concept car on display and you think, "I'd buy that in a second." And then five years later, the car finally comes to market and it's been whittled down from a Ferrari to a Pinto—all the truly breakthrough features have been toned down or eliminated altogether, and what's left looks mostly like last year's model. The same sorry fate could have befallen the iPod as well: Ive and Jobs could have sketched out a brilliant, revolutionary music player and then two years later released a dud. What kept the spark alive?

The answer is that Apple's development cycle looks more like a coffeehouse than an assembly line. The traditional way to build a product like the iPod is to follow a linear chain of expertise. The designers come up with a basic look and feature set and then pass it on to the engineers, who figure out how to actually make it work. And then it gets passed along to the manufacturing folks, who figure out how to build it in large numbers—after which it gets sent to the marketing and sales people, who figure out how to persuade people to buy it. This model is so ubiquitous because it performs

well in situations where efficiency is key, but it tends to have disastrous effects on creativity, because the original idea gets chipped
away at each step in the chain. The engineering team takes a look
at the original design and says, "Well, we can't really do that—but
we can do 80 percent of what you want." And then the manufacturing team says, "Sure, we can do some of that." In the end, the
original design has been watered down beyond recognition.

Apple's approach, by contrast, is messier and more chaotic at
the beginning, but it avoids this chronic problem of good ideas
being hollowed out as they progress through the development chain.
Apple calls it concurrent or parallel production. All the groups—
design, manufacturing, engineering, sales—meet continuously
through the product-development cycle, brainstorming, trading
ideas and solutions, strategizing over the most pressing issues, and
generally keeping the conversation open to a diverse group of perspectives. The process is noisy and involves far more open-ended
and contentious meetings than traditional production cycles—and
far more dialogue between people versed in different disciplines,
with all the translation difficulties that creates. But the results
speak for themselves.

Many of history's great innovators managed to build a cross-
disciplinary coffeehouse environment within their own private work routines. It is an oft-told story that Darwin delayed
publishing his theory of evolution because he feared the controversy it would unleash, particularly after the death of his beloved
daughter Annie traumatized his religious wife, Emma. But Darwin
also had an immense number of side interests to distract him from

his opus: he studied coral reefs, bred pigeons, performed elaborate taxonomical studies of beetles and barnacles, wrote important papers on the geology of South America, spent years researching the impact of earthworms on the soil. None of these passions were central to the argument that would eventually be published as *On the Origin of Species*, but each contributed useful links of association and expertise to the problem of evolution. The same eclectic pattern appears in countless other biographies. Joseph Priestley bounced between chemistry, physics, theology, and political theory. Even in the years before he became a political statesman, Benjamin Franklin conducted electricity experiments, theorized the existence of the Gulf Stream, designed stoves, and of course made a small fortune as a printer. While John Snow was solving the mystery of cholera in the streets of London in the 1850s, he was also inventing state-of-the-art technology for the administration of ether, publishing research on lead poisoning and the resuscitation of stillborn children, yet all the while tending to his patients as a general practitioner. Legendary innovators like Franklin, Snow, and Darwin all possess some common intellectual qualities—a certain quickness of mind, unbounded curiosity—but they also share one other defining attribute. They have a lot of hobbies.

The historian Howard Gruber likes to call such concurrent projects "networks of enterprise," but I prefer to describe them using a contemporary term that has been much maligned of late: multitasking. This is not, of course, the multitasking of the modern computer screen: switching from e-mail to spreadsheet to Twitter in a matter of seconds. What I'm describing is much more leisurely than that frenetic, digital-age mode; the individual tasks themselves might linger on for days or weeks before giving way to the next project. But there is steady variation nonetheless, not just in the subject matter but in the kind of work performed in each task. For John Snow, there were fundamentally different modes of intellectual activity involved in his many projects: building mechanical contraptions to control the temperature of chloroform required different skills and a different mind-set from tending to patients or writing papers for *The Lancet.* It is tempting to call this mode of work "serial tasking," in the sense that the projects rotate one after the other, but emphasizing the serial nature of the work obscures one crucial aspect of this mental environment: in a slow multitasking mode, one project takes center stage for a series of hours or days, yet the other projects linger in the margins of consciousness throughout. That cognitive overlap is what makes this mode so innovative. The current project can exapt ideas from the projects at the margins, make new connections. It is not so much a question of thinking outside the box, as it is allowing the mind to move through multiple boxes. That movement from box to box forces the mind to approach intellectual roadblocks from new angles, or to borrow tools from one discipline to solve problems in another.

The standard story about Snow is that he solved the mystery of cholera's waterborne transmission by doing shoe-leather epidemiological detective work during the 1854 Soho outbreak, but the truth is he had built a convincing rendition of the waterborne theory well before 1854. One reason he was able to see around the biases of the reigning "miasma" theory of the day—which maintained that cholera was caused by the inhalation of noxious vapors—is that his work with anesthesia had given him a hands-on knowledge of the way that gases diffused through the atmosphere. Snow reasoned that a disease transmitted by poisonous gas would leave a distinct pattern in the geographic spread of mortality: massive death in the immediate proximity of the bad smells, tapering off very quickly as one moved away from the original source. By the same token, Snow's training as a physician helped him shed the miasma blinders as well: from tending to patients ill with cholera, Snow observed that the effects of the disease on the human body indicated that the agent had been ingested, not inhaled, given that it did almost all of its direct damage in the digestive system and left the lungs largely unaffected. In a real sense, for Snow to make his great breakthrough in understanding cholera, he had to think like a molecular chemist *and* like a physician. As a slow multitasker, he had those interpretative systems readily available to him when his focus turned to the mystery of cholera. As we saw with the feathers of *Archaeopteryx*, Snow couldn't have anticipated that his mechanical tinkering with chloroform inhalers would prove useful in ridding the modern world of a deadly bacterium, but that is the unpredictable power of exaptations. Chance favors the connected mind.

VII.

PLATFORMS

On April 12, 1836, HMS *Beagle* took leave of the Keeling Islands, after a two-week idyll that had given Darwin the crucial evidence he needed to support the first great idea of his young career. As the ship left the placid green waters of the lagoon, heading home to England via the island of Mauritius, Captain FitzRoy plumbed the depths on the periphery of the atoll with a line more than 7,000 feet long. He encountered no bottom. FitzRoy's measurements confirmed, in Darwin's words, that the "island forms a lofty submarine mountain, with sides steeper even than those of the most abrupt volcanic cone." The data was crucial to Darwin, because he was building a theory in his mind about "lofty submarine mountains" and their geological legacy.

The theory had emerged years before as a hunch: that his mentor Charles Lyell's theory of atoll formation had a critical flaw that revolved around the statistical likelihood that a mountain would just happen to settle only a few feet above sea level. The elevation variation in volcanic islands was immense: some tapered off a dozen feet above sea level; others, like Mauna Kea, surged ten thousand feet into the sky. Most volcanic peaks lay thousands of feet below the surface. Yet Darwin, like most geologists of his age, knew that the oceans were populated by a huge number of tropical atolls that had all somehow simultaneously landed within a few feet of sea level. It was like scattering a hundred footballs across a field and having twenty of them cluster exactly on the forty-three-yard line. Darwin didn't have the theory of plate tectonics, but he knew that landmasses were rising and descending around the world. But it made no sense that these epic forces were somehow being arrested, in a significant number of cases, by the dividing line of sea level. A volcano being pushed upward by immense planetary conveyor belts should, by all rights, quickly burst through the ocean's surface and continue climbing, as Mauna Kea and countless other island volcanoes did. By the same logic, a mountain sliding into the sea should keep sliding. Why were so many of these mountains getting stuck?

We don't know exactly when the answer came to Darwin. It may well have occurred to him standing on the white sands of a Keeling Islands beach. More likely, knowing Darwin, the idea rolled in slowly, inch by inch, and some small piece of it came to him standing in those green waters. The idea was simple, but strangely hard to visualize. It began with one defining principle: the ground beneath Darwin's feet was not the product of geological forces. An organism had engineered it.

That organism was Scleractinia, more commonly known as reef-building coral. Alive, an individual Scleractinia is a soft polyp, no more than a few millimeters long. Reef-building corals grow in vast colonies, with new polyps appearing as buds on the sides of their "parents." It is one of the strange ironies of marine biology that the coral's essential contribution to the undersea ecosystem takes place after its death. The polyp builds a calcium-based exoskeleton during its life, producing a mineral called aragonite, which is sturdy enough to remain intact centuries after its original host has perished. A coral reef, then, is a kind of vast underwater mausoleum: millions of skeletons united to form the pocked, labyrinthine sprawl of a reef.

During his fortnight on the Keeling Islands, Darwin had observed that the soil of the island was entirely devoid of traditional rocks. As he put it in his diary, "Throughout the whole group of Islands, every single atom, even from the most minute particle to large fragments of rocks, bear [sic] the stamp of once having been subjected to the power of organic arrangement." The vast majority of those particles and rocks were aragonite skeletons, the remains of a coral polyp that had died decades or centuries before. This alone was evidence that Lyell's theory was flawed: if Darwin was standing at the tip of a dormant undersea volcano, the rocks at his feet would have been basalt or obsidian or pumice, rocks created from the cooling of molten lava. The rocks would have been forged in a fiery core of magma, not excreted by minuscule polyps.

The fact that the soil of an Indian Ocean atoll was organic in nature, engineered by coral and not the product of volcanic activity, did not, on its own, offer a satisfactory answer to the mystery of the atoll's existence. Why should a colony of coral form such a perfect oval in the middle of an immense ocean, hundreds of miles from another landmass? To solve that mystery, Darwin drew on Lyell's original theory, but he added an essential twist. He turned a still frame into a moving picture. To understand atoll formation, Darwin realized, you had to imagine a volcanic island slowly subsiding into the sea. As the banks of the volcano disappeared beneath the ocean waves, those slopes would become prime breeding ground for coral colonies, which thrive in shallow water at depths up to around 150 feet. (Their diet relies substantially on photosynthetic algae that cannot survive too far from the sunlit surface of the water.) Eventually the summit of the mountain slides into the sea, leaving a circle of shallow water defined by the periphery of the volcanic crater. Because the mountain is subsiding so slowly, the coral are able to build their reefs faster than the mountain can descend. Like overzealous developers, the coral colonies keep adding new floors to the structure they've erected at the top of the volcano, limited only by the water's surface. As the original peak descends further and further into the sea, the older reefs die off, but continue to give structural support to the new, thriving reefs above them. Darwin had no way of measuring this precisely, but he predicted that fossil coral would extend as far as five thousand feet below sea level before hitting a volcanic foundation, a number that was confirmed more than a century later with modern drilling technology.

As the *Beagle* departed, Darwin captured the miraculous nature of this explanation in his diary. "We must look at a Lagoon [island] as a monument raised by myriads of tiny architects," he wrote, "to mark the spot where a former land lies buried in the depths of the ocean."

Published several years later as a monograph, Darwin's theory of atoll formation marked his first significant contribution to science, and it has largely stood the test of time. The idea itself drew on a coffeehouse of different disciplines: to solve the mystery, he

had to think like a naturalist, a marine biologist, and a geologist all
at once. He had to understand the life cycle of coral colonies, and
observe the tiny evidence of organic sculpture on the rocks of the
Keeling Islands; he had to think on the immense time scales of
volcanic mountains rising and falling into the sea. And, of course,
he needed FitzRoy's technical expertise with the sounding line. To
understand the idea in its full complexity required a kind of probing intelligence, willing to think across those different disciplines
and scales. Darwin described it best in the chapter on his Keeling
Islands investigations from *The Voyage of The Beagle*: "We feel
surprise when travelers tell us of the vast dimensions of the Pyramids and other great ruins, but how utterly insignificant are the
greatest of these, when compared to these mountains of stone accumulated by the agency of various minute and tender animals. This
is a wonder which does not at first strike the eye of the body, but,
after reflection, the eye of reason."

From Darwin's perspective, those "minute and tender animals"
had built a platform, in the most prosaic sense of the word. Darwin
was walking on that saucer-shaped summit, and not treading water
in the middle of the Indian Ocean, because those animals had engineered a platform for him to stand on. But a coral reef is a platform
in a much more profound sense: the mounds, plates, and crevices of
the reef create a habitat for millions of other species, an undersea
metropolis of immense diversity. To date, attempts to measure accurately the full diversity of reef ecosystems have been foiled by
the complexity of these habitats; scientists now believe that somewhere between a million and ten million distinct species live in coral
reefs around the world, despite the fact that those reefs only occupy
one-tenth of one percent of the planet's surface. This is the Darwin

Paradox: that such nutrient-poor waters could generate so much marvelous, improbable, heterogeneous life.

For forty years, ecologists have used the term "keystone species" to designate an organism that has a disproportionate impact on its ecosystem—a carnivore, for instance, who is the only predator of another species that would otherwise overwhelm the habitat with unchecked population growth. Remove the keystone predator and the habitat falls apart. But about twenty years ago, a scientist named Clive Jones at the Cary Institute of Ecosystem Studies decided that ecology needed another term to describe a very specific kind of keystone species: the kind that actually creates the habitat itself. Jones called these organisms "ecosystem engineers." Beavers are the classic example of ecosystem engineers. By felling poplars and willows to build dams, beavers single-handedly transform temperate forests into wetlands, which then attract and support a remarkable array of neighbors: pileated woodpeckers drilling nesting cavities into dead trees; wood ducks and Canada geese settling in abandoned beaver lodges; herons and kingfishers and swallows enjoying the benefits of the "artificial" pond, along with frogs, lizards, and other slow-water species like dragonflies, mussels, and aquatic beetles. As do those underwater colonies of coral, the beaver creates a platform that sustains an amazingly diverse assemblage of life.

Platform building is, by definition, a kind of exercise in emergent behavior. The tiny Scleractinia polyp isn't actively trying to create an underwater Las Vegas, but nonetheless out of its steady labor—imbibing algae and erecting those aragonite skeletons—a higher-level system emerges. What had been a largely desolate stretch of nutrient-poor seawater is transformed into a glittering hub of activity. The beaver builds a dam to better protect itself against its predators, but that engineering has the emergent effect of creating a space where kingfishers and dragonflies and beetles can make a life for themselves. The platform builders and ecosystem engineers do not just open a door in the adjacent possible. They build an entire new floor.

The cafeteria at the Johns Hopkins University Applied Physics Laboratory in Laurel, Maryland, had long been a site of productive shoptalk between the physicists, technicians, mathematicians, and proto-hackers who worked there. But the Monday lunchtime chatter on October 7, 1957, was unusually heated, thanks to the weekend headlines announcing the Soviet launch of *Sputnik 1*, the first man-made earth-orbiting satellite. Two young physicists, William Guier and George Weiffenbach, found themselves in a spirited discussion about the microwave signals that would likely be emanating from *Sputnik*. After canvassing some of their colleagues, it appeared that no one had bothered to come in over the weekend to see if *Sputnik*'s signals could be picked up by the APL's equipment. Weiffenbach, as it turned out, was in the middle of a Ph.D. on microwave spectroscopy and had a 20 MHz receiver sitting in his office.

Guier and Weiffenbach spent the afternoon hunched over the receiver, listening for *Sputnik*'s audio fingerprint. To combat the doubters, who would inevitably question whether the whole launch was an elaborate hoax, a product of communist propaganda, the Soviets had engineered *Sputnik* so that it would transmit an unusually accessible signal: an unbroken tone broadcast within 1 kHz of 20 MHz. By the end of the afternoon, Weiffenbach and Guier had a clear lock on it. The sound itself was a staccato pulse of electronic bleeps, but the context transformed it into the most marvelous music the two men had ever heard. It seemed unbelievable: sitting in a room in suburban Maryland, listening to man-made signals coming from space. Word began to spread through the APL that the young physicists had captured *Sputnik*'s signal, and a steady stream of visitors appeared at Weiffenbach's door to eavesdrop on the satellite's warble.

Realizing that they were listening to history, Guier and Weiffenbach hooked up the receiver to an audio amplifier and began recording the signal on audiotape. They included time stamps with each recording. As they listened and recorded, the two men realized that they could use the Doppler effect to calculate the speed at which the satellite was moving through space. First observed more than a century before by the Austrian physicist Christian Doppler, the Doppler effect describes the predictable way a waveform's frequency changes when the source or the receiver is in motion. Imagine a speaker playing a single note, let's say the A above middle C, which sends out sound waves with a frequency of 440 Hz. If you mount the speaker on the hood of a car and have it driven toward you, the waves stack up on top of each other, making the interval between each of them shorter. When those compressed waves arrive in your eardrum, their perceived frequency is higher than 440 Hz. When the car backs up, the Doppler effect reverses, and the perceived note drops below A. You can hear the Doppler effect at work every time an ambulance drives past you with blaring sirens; as it passes you by, the sound of its siren appears to slide down in pitch.

The Doppler effect has proved to be a remarkably versatile concept: it has been used to detect the expansion of the universe, to track thunderstorms, and to perform ultrasounds. Because *Sputnik*

was emitting a signal at a steady frequency, and because the microwave receiver was stationary, Guier and Weiffenbach realized that
they could calculate the movement of the satellite based on the
small but steady changes in the waveform they were capturing. Late
that night, they remembered an additional mathematical trick: by
analyzing the slope of the Doppler shift, they could determine the
point in *Sputnik*'s orbit that was closest to the APL laboratories. Almost by accident, they had hit upon a technique not just for calculating the satellite's speed, but for actually mapping the trajectory
of its orbit. In a matter of hours, the two young scientists had gone
from listening to measuring to tracking the Russian satellite.

Over the subsequent weeks, a loose network of scientists at
APL coalesced around Guier and Weiffenbach's hunch, filling in
details, researching the theoretical literature on orbiting bodies, and
proposing technology improvements. Eventually, the APL's director
approved funds to run the numbers on the lab's new UNIVAC computer. Within a few months of that first transmission, they had a
complete description of *Sputnik*'s orbit, inferred entirely from that
simple 20 MHz signal. Guier and Weiffenbach had embarked on a
quest that would define their professional careers, the "adventure
of their lives," as they later called it. In the spring of 1958, Frank
T. McClure, the legendary deputy director of the Applied Physics
Laboratory, called Guier and Weiffenbach into his office. McLure
had a confidential question to ask the men: If you could use the
known location of a receiver on the ground to calculate the location
of a satellite, McClure asked, could you reverse the problem? Could
you calculate the location of a receiver on the ground if you knew
the exact orbit of the satellite? Guier and Weiffenbach ran the logic
through their heads for a few minutes, and then answered in the

affirmative. In fact, deducing the location from a known orbit—instead of a stationary ground position—would make the results significantly more accurate. Without explaining his ultimate interest in the question, McClure told the two men to run a quick feasibility analysis. After a few furious days of crunching the numbers, Guier and Weiffenbach reported back: the "inverse problem," as they called it, was eminently solvable.

Soon, Guier and Weiffenbach would learn why the inverse problem was so important to McClure: the military was developing its Polaris nuclear missiles, designed to be launched from submarines. Calculating accurate trajectories for a missile attack required precise knowledge of the launch site's location. This was easy enough to determine on land—say, for a missile silo in Alaska—but it was fiendishly difficult in the case of a submarine floating somewhere in the Pacific Ocean. McClure's idea was to take the ingenious *Sputnik* solution and flip it on its head. The military would establish the unknown location of its submarines by tracking the known location of satellites orbiting above the earth. Just as sailors had used the stars to navigate for thousands of years, the military would steer its ships using the artificial stars of satellite technology.

The project was dubbed the Transit system. Just three years after *Sputnik*'s launch, there were five U.S. satellites in orbit, providing navigational data to the military. When Korean Air Lines Flight 007 was shot down in 1983 after drifting into Soviet airspace thanks to faulty, ground-based navigation beacons, Ronald Reagan declared that satellite-based navigation should be a "common good" open to civilian use. Around that time, the system took on its current name: Global Positioning System, or GPS. Half a century later, roughly thirty GPS satellites blanket the earth with navigational signals, providing guidance for everything from mobile phones to digital cameras to Airbus A380s.

If you wish to see firsthand the unpredictable power of an emergent platform, you need only look at what has happened to GPS over the past five years. The engineers that built the system—starting with Guier and Weiffenbach—created an entire ecosystem of unexpected utility. Frank McClure recognized that you could harness Guier and Weiffenbach's original insight to track nuclear submarines, but he had no inkling that fifty years later the same system would help teenagers to play elaborate games in urban centers, or climbers to explore treacherous mountain ranges, or photographers to upload their photos to Flickr maps. Like the Internet itself, GPS has turned out to have immense commercial value, and many for-profit firms were involved in building out the infrastructure that made it a reality. But the *ideas* at the foundation of GPS—the notion of a satellite itself, the atomic clocks satellites rely on for accurate timing, and, of course, Guier and Weiffenbach's original insight with *Sputnik*—all came out of the public sector. The generative nature of the GPS platform nicely mirrors the original environment that gave birth to it. When Guier and Weiffenbach were asked to explain how they had hit upon their *Sputnik* revelation, they credited the intellectual habitat of the Applied Physics Lab more than their own particular talents:

> APL was a superb environment for inquisitive young kids, and particularly so in the Research Center. It was an environment that encouraged people to think broadly and generally about task problems, and one in which inquisitive kids felt free to follow their curiosity. Equally important, it was an environment

wherein kids, with an initial success, could turn to colleagues
who were broadly expert in relevant fields, and particularly because of the genius of the Laboratory Directorship, colleagues
who were also knowledgeable about hardware, weapons, and
weapons needs.

In its own small way, the APL was a platform that encouraged and
amplified hunches, that allowed those hunches to be connected with
other minds that had relevant expertise. Out of that dense network,
one of the most generative technological platforms of the twenty-
first century took root. The APL was not a purely open platform, of
course. There were military secrets involved, after all; and even if
Guier and Weiffenbach had wanted to share their *Sputnik* discovery
with the world, it was much harder to distribute that breakthrough
in an age when the hot new computer—the UNIVAC—took up an
entire room. But behind those closed doors, William Guier and
George Weiffenbach were the beneficiaries of an environment that
encouraged the chance collisions between different fields, an environment that let two "kids" stumble across an idea at the cafeteria
and build an entire career around it.

Most hotbeds of innovation have similar physical spaces associated with them: the Homebrew Computing Club in Silicon Valley; Freud's Wednesday salon at 19 Berggasse; the eighteenth-century
English coffeehouse. All these spaces were, in their own smaller-
scale fashion, emergent platforms. Coffeehouse proprietors like Edward Lloyd or William Unwin were not trying to invent the modern
publishing industry or the insurance business; they weren't at all
interested in fostering scientific advancement or political turmoil.

They were just businessmen, trying to make enough sterling to feed their families, just like those beavers constructing lodges to keep their offspring safe. But the spaces Lloyd and Unwin built turned out to have these unusual properties: they made people think differently, because they created an environment where different kinds of thoughts could productively collide and recombine.

The most generative platforms come in stacks, most conspicuously in the layered platform of the Web. (The phrase "platform stack" itself is part of the common parlance of modern programming.) The Web can be imagined as a kind of archaeological site, with layers upon layers of platforms buried beneath every page. Tim Berners-Lee was able to single-handedly design a new medium because he could freely build on top of the open protocols of the Internet platform. He didn't have to engineer an entire system for communicating between computers spread across the planet; that problem had been solved decades before. All he had to do was build a standard framework for describing hypertext pages (HTML) and sharing them via existing Internet channels (HTTP). Even HTML was based on another existing platform, SGML, which had been developed at IBM in the 1960s. Fourteen years later, when Hurley, Chen, and Karim sat down to create YouTube, they built the service by stitching together elements from three different platforms: the Web itself, of course, but also Adobe's Flash platform, which handled all the video playback, and the programming language Javascript, which allowed end users to embed video clips on their own sites. Their ability to build on top of these existing platforms explains why three guys could build YouTube in six months, while an army of expert committees and electronics companies took twenty years to make HDTV a reality.

Culture, too, relies on stacked platforms of information. Kuhn's paradigms of research are the scientific world's equivalent of a software platform: a set of rules and conventions that govern the definition of terms, the collection of data, and the boundaries of inquiry for a particular field. Kuhn's argument has often been mistaken as a defense of a purely relativistic account of science, where empirical "truth" is always in quotation marks because paradigms replace each other over time. (The apparent solidity of scientific truth, in this account, is merely a kind of hologram produced by the apparatus of the paradigm.) But modern scientific paradigms are rarely overthrown. Instead, they are built upon. They create a platform that supports new paradigms above them. Darwin's theory of natural selection was a "dangerous" idea—in Daniel Dennett's phrase—because it challenged Biblical and human-centric accounts of life's history, but the true measure of its scientific power lies in how many new fields were stacked on top of it over the course of the twentieth century: the Mendelian and population genetics that emerged from the "modern synthesis" in the 1940s; the molecular genetics revolution triggered by Watson and Crick's discovery of DNA; newer fields like evolutionary psychology and "evolutionary development." Often, new scientific fields form by propping themselves over *multiple* platforms. The field that ultimately explained Darwin's Paradox—ecosystems ecology—stands on the shoulders of population genetics, systems theory, and biochemistry, among others.

Even the creative arts evolve via stacked platforms. This may seem surprising, given how readily we draw upon the image of the private artistic genius, holed up in his study, conjuring a whole new world in his head from scratch. For understandable reasons, we like to talk about artistic innovations in terms of the way that they break the rules, open up new doors in the adjacent possible that lesser minds never even see. But genius requires genres. Flaubert and Joyce needed the genre of the bildungsroman to contort and undermine in *Sentimental Education* and *A Portrait of the Artist as a Young Man*. Dylan needed the conventions of acoustic folk to electrify the world with *Highway 61 Revisited*. Genres supply a set of implicit rules that have enough coherence that traditionalists can safely play inside them, and more adventurous artists can confound our expectations by playing *with* them. Genres are the platforms and paradigms of the creative world. They are almost never willed into existence by a single pioneering work. Instead, they fade into view, through a complicated set of shared signals passed between artists, each contributing different elements to the mix. The murder mystery has been coherent as a novelistic genre for a hundred years, but when you actually chart its pedigree, it gets difficult to point to a single donor: it's a little Poe, a little Dickens, a little Wilkie Collins, not to mention the dozens of contemporaries who didn't make the canon, but who nonetheless played a role in stabilizing the conventions of the genre. The same is true of cubism, the sitcom, romantic poetry, jazz, magical realism, cinema verité, adventure novels, reality TV, and just about any artistic genre or mode that has ever mattered.

The creative stack is deeper than genres, though. Genres are themselves built on top of more stable conventions and technologies. When Miles Davis announced his break with the chord-and-improv conventions of bebop jazz in "So What?"—the opening track of *Kind of Blue*—he was nonetheless working within the conventions of the D Dorian scale the song employs, a mode that, as its name suggests, dates back to the Dorian Greeks. And of course, Davis built his new sound out of the older, stable platforms of the instruments themselves, starting with the valved trumpet that Davis played. "Natural" trumpets—lacking the complex valves that allow the trumpeter to switch keys on the fly—are almost as old as the Dorian mode; the modern valved trump that Davis played emerged as a standard in the nineteenth century, after decades of tinkering by instrument makers across Europe. Davis could afford to explore the adjacent possible of jazz, to help invent a whole new genre that others would build upon, in part because he didn't have to invent the D Dorian or the valved trumpet.

In the online world, the most celebrated recent case study in the innovative power of stacked platforms has been the rapid evolution of the social networking service Twitter. Twitter's creators, Jack Dorsey, Evan Williams, and Biz Stone, benefited from existing platforms just as the YouTube founders did: Twitter's legendary 140-character limit is based on the limitations of the SMS mobile communications platform that they rely on to connect Web messages to mobile phones. But the most fascinating thing about Twitter is how much has been built on top of its platform in three short years. When it first emerged, Twitter was widely derided as a frivolous distraction that was mostly good for telling your friends what you had for breakfast. Now it is being used to organize and share news about the Iranian political protests, to route around government censorship, to provide customer support for large corporations, to share interesting news items, and a thousand other applications that did not occur to the founders when they dreamed up the service in 2006. This is not just a case of cultural exaptation: people finding a new use for a tool designed to do something else. In Twitter's case, the users have been redesigning the tool itself. The convention of replying to another user with the @ symbol was spontaneously invented by the Twitter user base. Early Twitter users ported over a convention from the IRC messaging platform, and began grouping a topic or event by the "hash-tag," as in "#30Rock" or "#inauguration." The ability to search a live stream of tweets—which is likely to prove crucial to Twitter's ultimate business model, thanks to its advertising potential—was developed by another start-up altogether. Thanks to these innovations, following a live feed of tweets about an event—political debates or *Lost* episodes—has become a central part of the Twitter experience. But for the first year of Twitter's existence, that mode of interaction would have been technically impossible using Twitter. It's like inventing a toaster oven and then looking around a year later and discovering that all your customers have, on their own, figured out a way to turn it into a microwave.

One of the most telling facts about the Twitter platform is that the vast majority of its users interact with the service via software that has been created by third parties. There are hundreds of iPhone and BlackBerry applications that let you manage your Twitter feeds, all created by enterprising amateur coders or small start-ups. There are services that help you upload photos and link to them from your tweets; programs that map other Twitizens who are near you geographically. Ironically, the tools you're offered if you visit the Twitter.com site have changed very little in the past two years. But there's an entire Home Depot of Twitter tools available everywhere else.

The diversity of the Twitter platform is no accident. It derives from a deliberate strategy that Dorsey, Williams, and Stone embraced from the outset: they built an emergent platform first, and *then* they built Twitter.com. An open platform in software is often called an API, which stands for application programming interface. An API is a kind of lingua franca that software applications can use reliably to communicate with each other, a set of standardized rules and definitions that allow programmers to build new tools on top of another platform, or to weave together information from multiple platforms. When Web users make geographic mashups using Google Maps, they write programs that communicate with Google's geographic data using their mapping API.

Some APIs reveal only a small subset of a platform's underlying code, in part for simplicity's sake, but also for proprietary reasons. Conventionally, a developer will create a piece of software, and once it's finished, expose a small part of its functionality to outside developers via the API. The Twitter team took the exact opposite approach. They built the API first, and exposed all the data that was crucial to the service, *and then they built Twitter.com on top of the API.* Conventional software assumes that API users are second-class citizens who shouldn't get full access to the software's secret sauce for fear of losing competitive advantage. Twitter's creators recognized that there was another kind of competitive advantage that came from complete openness: the advantage that comes from having the largest and most diverse ecosystem of software applications being built on your platform. Call it cooperative advantage. The burden of coming up with good ideas for the product is no longer shouldered exclusively by the company itself. On an open platform, good ideas can come from anywhere.

The way for-profit companies like Twitter and Google have used open APIs to spur innovation has been fascinating to watch.

But the more intriguing developments lie in the public sector. In the fall of 2008, Vivek Kundra, chief technology officer for the District of Columbia, announced a program called Apps for Democracy (replacing its somewhat more scandalous working title, Hack the District). Software developers were invited to build applications that drew upon the open data made available by the city government. The applications could take just about any form imaginable—websites, Facebook applications, iPhone apps—as long as they attempted to make some part of the government data trove more useful for residents, visitors, businesses, or government agencies. The winners would receive $10,000 in prize money.

The city provided just thirty days for developers to create their applications, but even in that narrow window, forty-seven different applications were submitted. The two winning applications showcased historic walking tours around the D.C. area and provided extensive demographic information for residents thinking of moving to a new neighborhood. Other submissions included tools for tracking government spending on specific projects, guides for city bikers, and real-time parking information with data received directly from on-street parking meters. One ingenious, and amusing, app, called StumbleSafely, helped inebriated users to plot the safest pedestrian route home from any bar in the city.

The D.C. experiment was such a success that versions of it are currently proliferating in dozens of major cities around the world. When D.C. launched its second iteration of Apps for Democracy, in the spring of 2009, Kundra wasn't around to award the prizes, but for a good reason: he had been appointed the nation's chief information officer by President Obama, helping to create the ambitious Data.gov program, along with an Apps for America contest run by the Sunlight Foundation. What these initiatives share is a willingness to learn from the innovation platforms of Twitter, Google, and Facebook. When Al Gore set about to "reinvent government" during the Clinton administration, one of that project's ambitious goals was to make the bureaucracy more innovative. But Gore's solutions were, almost without exception, inward-facing: creating new organizational structures inside the government, cutting down on red tape; encouraging cross-departmental collaboration. What Apps for Democracy suggests is a more open-ended idea: some of the best ideas for government are likely to come from *outside* the government. If the outside-developer community could build something as essential to Twitter's business as a search interface, then why can't citizen developers provide comparable innovations for their government? Surely someone out there can come up with a better user experience for filing tax forms.

Government bureaucracies have a long and richly deserved reputation for squelching innovation, but they possess four key elements that may allow them to benefit from the innovation engine of an emergent platform. First, they are repositories of a vast amount of information and services that could be of potential value to ordinary people, if only we could organize it all better. Second, ordinary people have a passionate interest in the kind of information governments deal with, whether it's data about industrial zoning, health-care services, or crime rates. Third, a long tradition exists of citizens committing time and intellectual energy to tackling problems where there is a perceived civic good at stake. And, finally, the fact that governments are not in the private sector means that they do not feel any competitive pressure to keep their data proprietary.

Since the supernova that was the Howard Dean campaign in 2004, it has been clear that network technology can be harnessed to help our leaders run for office. But we have not yet seen real evidence that these extraordinary technologies can help those leaders govern more effectively once they get elected. But thinking of government as a platform—to borrow a phrase from Web visionary Tim O'Reilly—might be one way to carry out the promise of digital-age governance. Political leadership involves some elements that aren't best outsourced to a liquid network; decision-making and oratory. But a good government is, at least in part, a government that comes up with innovative solutions to the problems of its citizens, or to the problems faced by bureaucracy itself. That's where the platform model can do its magic.

Part of that magic is economic: emergent platforms can dramatically reduce the costs of creation. Those forty-seven apps generated in a month by the original Apps for Democracy contest had a total cost to the D.C. government of $50,000. Kundra estimated that, had the city government contracted out for those applications using its traditional methods, the cost to the city would have been more than $2,000,000. (Also, the process would have taken more than a year.) The same math applies to private-sector Web innovation. If Hurley, Chen, and Karim had been forced to concoct an online video standard from scratch, it would have taken years and tens of millions of dollars just to get a working beta version online. To this day, Twitter has not spent a dime building a mapping application to track the location of tweets, because dozens of services exist that do exactly that, created and promoted by third parties at zero cost to Twitter itself.

Though they are not measured in monetary units, natural plat-

forms display similar patterns of economic efficiency. Pileated
woodpeckers make their homes by drilling large holes in dead trees.
But woodpeckers don't have the resources to kill off trees on their
own, so they're largely dependent on stumbling across trees that
have died of natural causes. But in creating their forest wetlands,
beavers are constantly toppling trees, and so pileated woodpeckers
flourish in the engineered ecosystem created by the beavers. They
get the benefit of the softer, more pliable wood of a rotting tree,
without the cost of having to fell the tree. Interestingly, woodpeckers
generally abandon the homes they've carved into the tree after a
year, making them ideal spaces for songbirds to nest. The songbirds
benefit from the cavities created by the woodpeckers without being
burdened by the costs of drilling through all that wood. The wetland
created by the beaver, like the thriving platform created by the Twitter founders, invites variation because it is an open platform where
resources are shared as much as they are protected.

If you sail due east sixteen nautical miles from Delaware's Indian
River Inlet, and dive eighty feet down into the open waters
of the Atlantic Ocean, you will discover an underwater city thriving on the seafloor: massive schools of flounder, sea bass, and tau-
tog darting through gently waving sea grasses. You will also find
roughly seven hundred subway cars, deposited there by the Delaware Department of Natural Resources and Environmental Control
over the past decade. The trains have been planted off the Delaware
shore to create an artificial reef, providing a durable shelter for
mussels and sponges that are otherwise challenged by the sandy

floors of the northeast seaboard. Artificial reefs create significant breeding grounds for a diverse group of fish; the Delaware reefs have seen a 400 percent increase in biomass since the first cars were sunk. (Artificial reefs also have the secondary effect of preventing beach erosion.) No longer needed for mass transportation, the abandoned subway cars have taken on a new occupation in their retirement years. They are now ecosystem engineers.

Platforms have a natural appetite for trash, waste, and abandoned goods. The sea bass and mussels making a home in a decommissioned A train, like the songbirds nesting in the abandoned homes of the pileated woodpeckers, mirror a pattern Jane Jacobs detected years ago in urban development: innovation thrives in discarded spaces. Emergent platforms derive much of their creativity from the inventive and economical reuse of existing resources, and, as any urbanite will tell you, the most expensive resource in a big city is real estate. "If you look about, you will see that only operations that are well established, high-turnover, standardized or heavily subsidized can afford, commonly, to carry the costs of new construction," Jacobs wrote. "Chain stores, chain restaurants and banks go into new construction. But neighborhood bars, foreign restaurants and pawn shops go into older buildings. Supermarkets and shoe stores often go into new buildings; good bookstores and antiques dealers seldom do." One implication of this is that riskier or smaller-scale enterprises tend to have difficulty getting traction in planned environments that lack the economic wear and tear of the traditional urban fabric, where buildings, blocks, and whole neighborhoods lose their original inhabitants and industries, sometimes to catastrophic effects. (The closest suburban approximation is the marginal space of the garage, where Hewlett-Packard, Apple, and Google all set their original roots.) The shopping mall is only fifty years old, and so is relatively young by the millennial scale of some cities, but thus far even the most down-on-its-luck mall has retained its original function: as a place where consumers gather to buy things for personal use. They have not yet been reclaimed by troupes of performance artists, or Internet start-ups, or heavy industry. There are streets in the West Village of Manhattan, where Jacobs lived for so many years, that now resemble shopping malls. But over the past two centuries those old buildings have hosted an entire cavalcade of different uses: they have served as the hub of an industrial port; as the primary supply point of meat for a city of eight million people; as a refuge for beatniks and dropouts; as the epicenter of the gay rights movement. Jacobs's point was that the frenetic energy of a large city, the urban version of creative destruction, creates a natural supply of older, less-desirable environments that can be imaginatively reoccupied by the small or the eccentric, the subcultures that Fischer found so essential to urban life. Artists, poets, and entrepreneurs are the vibrant fish swimming among the coral of the Keeling Islands: they find it easier to live in an exoskeleton that has long since been abandoned by its original host. As Jacobs observed:

> As for really new ideas of any kind—no matter how ultimately profitable or otherwise successful some of them might prove to be—there is no leeway for such chancy trial, error and experimentation in the high-overhead economy of new construction. Old ideas can sometimes use new buildings. New ideas must use old buildings.

Platforms recycle much more than just architecture. Marine ecologists who have studied the flow of energy through coral reef ecosystems have found that coral reefs do an astounding job of recycling nutrients. Scientists have long recognized the importance of the symbiotic relationship between the coral and a microscopic algae called zooxanthella. The two organisms effectively rely on each other's waste products: the algae captures energy from the sun and outputs oxygen and sugars as waste, which the coral polyps use to power their own growth. At the same time, the corals expel carbon dioxide, nitrates, and phosphates as waste, each of which fuels the growth of zooxanthellae. As the population of zooxanthellae expands, more solar energy is captured and thus available to be shared with the broader ecosystem of the reef. The zooxanthella and the coral are like two neighbors who miraculously turn out to have a pressing need for each other's garbage, and thus meet every night to swap trash cans.

But the nutrient recycling of a coral reef extends far beyond the collaboration between coral and zooxanthella. In 2001, a team of German ecologists led by Claudio Richter used endoscopes to examine the tiny internal cavities of coral reefs in the Red Sea. Hidden in those diminutive grottoes was a vast population of sponges that have adapted to the dark interior of the coral reef because it provides them sanctuary from their natural predators, sea urchins and parrotfish. The sponges consume another key photosynthetic organism, phytoplankton, as it drifts through the aragonite caves of the reef. Like the zooxanthellae, the sponges then expel waste products that the coral can use as nutrients. Those long-hidden sponges embody two principles of platform recycling: by co-opting the abandoned space of the coral skeleton, they reduce the costs of fortifying themselves against predators. And in return, they expel nutrients that allow their host to excrete even more aragonite, creating new habitats for more sponges.

The entire coral reef ecosystem is characterized by similarly intricate and interdependent food webs, the full complexity of which scientists are only now beginning to map. Once you understand the way biological platforms build on the waste products generated within the system, Darwin's Paradox ceases to be a paradox at all. The symbiotic relationship between coral and zooxanthella increases the total energy captured from the sun, and the tight nutrient cycles created by the productive reuse of energy sources by so many densely interconnected species means that the habitat can do much more with less. You get a watery metropolis with astonishing diversity in an environment that by rights should be as desolate as the sandy atoll above sea level. It is not competition that drives that process, but rather the inventive collaborations of density. The reef platform does not have the luxurious supply of nutrients that tidal estuaries do, delivered daily by the freshwater rivers that carve topsoil out of riverbanks upstream. But the reef platform thrives nonetheless, thanks to the ecosystem engineering of the coral, and the marvelous recycling of both shelter and biological waste that makes the platform so vital.[5] Above the waterline, on those vacant atolls, a markedly different landscape appears, much closer to the wasteful ecosystems of deserts. Most of the solar energy that saturates desert environments

---

5. The same pattern appears in rain forests, precisely because there are so many organisms exploiting every tiny niche of the nutrient cycle. That efficiency is one of the reasons that clearing the rain forests is such a shortsighted move: the nutrient cycles in rain-forest ecosystems are so tight that the soil is usually very poor for farming—all the available energy has been captured on the way down to the earth.

gets lost, assimilated only by the few succulents that can survive in such a hostile climate. Those plants pass along enough energy to sustain a limited number of insects, which in turn supply food for the occasional reptile or bird, all of which ultimately feeds the bacteria. But most of the energy never gets put to use by organic life.

If Brent Constantz has his way, the coral reef's genius for recycling and platform building will end up transforming the physical platforms of human settlements. In the late seventies, while pursuing a Darwinesque double major in biology and geology at the University of California, Santa Barbara, Constantz became fascinated by the coral polyp's extraordinary powers of biomineralization, its ability to build an immense structure of calcium carbonate durable enough to last for millions of years. Human beings may be justifiably proud of venerable engineering achievements like the Pyramids or the Great Wall of China, but those monuments pale in comparison to the Great Barrier Reef, the largest biological structure on the planet. As an undergrad, Constantz daydreamed about harnessing the coral's engineering skills to create entire buildings out of prefab templates. Instead of pouring concrete or attaching steel beams, the templates would simply be lowered into seawater, where the reef-building process would magically conjure up a building. It was a fantasy in those years, but Constantz kept that strange vision in the back of his mind for decades.

By 1985, Constantz was most of the way through a Ph.D. at U.C. Santa Cruz, and had become an expert in the techniques of biomineralization. On his way to a research expedition funded by a NSF grant, he stopped over for a few days to see his parents at their home near Palo Alto. Watching a football game with his father, a physician, he picked up a medical journal and stumbled across an article about the massive health expenses associated with osteoporosis, a disease that disrupts bone mineral density, causing painful and debilitating fractures. A few weeks later, he was standing on the Rangiroa atoll in the middle of the Pacific Ocean, measuring the speed with which the corals built their skeletons, and his mind flashed back to the osteoporosis article. "If you could somehow capture these skeleton growth processes," he thought, "you could really help all those old ladies with broken hips." Two year later, he started his first company, which mimicked the coral's growth mechanism to create bone cement to repair fractures. Today, the cements that Constantz created are employed in most orthopedic operating rooms throughout the United States and Europe.

Constantz went on to found two other successful biomedical companies, but that original hunch about building physical infrastructure out of coral skeletons lingered in the back of his mind. While teaching at Stanford in the mid-2000s, he joined the cross-disciplinary faculty of the Woods Institute for the Environment, where for the first time he learned about the mammoth environmental impact of manufacturing Portland cement, the third largest source of human-created carbon dioxide emissions on the planet. In his mind, a new network of ideas began to take shape, reviving his old undergraduate dream of growing aquatic cities. Coral reefs created cement-like structures without polluting the environment, and Constantz had three successful companies to show that mimicking the coral growth mechanics could create useful new materials. What if you took those mechanics and used them for building highway overpasses instead of repairing hip fractures?

The slow hunch he had been nurturing for twenty-five years had finally found the right connection. He took his vision of a "green" cement to one of Silicon Valley's legendary venture capitalists, Vinod Khosla, who agreed to fund the company (which Constantz named Calera) without seeing so much as a business plan or PowerPoint deck. Constantz built a laboratory in Los Gatos, where they began "growing" carbonate cement in transport trailers filled with seawater. He soon discovered that the system generated eight times as much cement if you pumped the water full of carbon dioxide, like some oversized, salty club soda. One day, when Khosla came to inspect the lab, Constantz turned to his investor and asked, "Where can we get large quantities of carbon dioxide?" Khosla looked at him in disbelief. As one of the world's most prominent clean-tech investors, Khosla was well aware that the planet was teeming with industrial plants who were desperately trying to find a place to put their carbon dioxide. Entire markets were emerging around technologies of carbon sequestration, locking up CO_2 by injecting it into oil and gas reserves, or burying it deep in the ocean. But Constantz had stumbled across a much more powerful idea. You didn't have to bury all that CO_2. You could use it to build stuff.

The Calera story is still very much in progress. It remains an open question whether the cities of our future will be built underwater by virtual coral reefs on a diet of factory exhaust. It sounds fanciful when described that way, of course, but no more fanciful than the idea of the Great Barrier Reef would have seemed a billion years ago. Nature has long built its platforms by recycling the available resources, including the waste generated by other organisms. Two things we have in abundance on this planet right now are pollution and seawater. Why not try to build a city out of them?

.   .   .

The stacked platform of the Web depends on recycling as well. The word "ecosystem" has become a fashionable term to describe the diverse collection of sites and services associated with Web 2.0. Like most jargon, the metaphor points to an important truth, if you think of the flows of information across the Web as being analogous to the flows of energy through a natural ecosystem. But also like most jargon, the metaphor is too general, and its broad scope actually makes it harder for us to see the most important thing about the evolution of the Web over the past fifteen years. The Web is not simply an ecosystem; it is a specific *type* of ecosystem. It started as a desert, and it has been steadily transforming into a coral reef.

Part of the beauty and power of Tim Berners-Lee's architecture for the Web lies in its simplicity: websites were made up of hypertext pages that could connect to other information on the Web through one primary conduit: the link. Imagine it's 1995, and you decide to post a short review of a new restaurant in Boston's Back Bay neighborhood on your "home page," as we used to call them back then. In posting that restaurant review, you are contributing new information to the Web's ecosystem. Like zooxanthellae capturing energy from the sun, you are taking information originally created outside the environment (in the neural networks of your own brain) and adding it to the information resources available on the Web.

The question is, what happened to that information once you added it to the system? You could link to the home page of the restaurant itself, if you were lucky enough to find one in those early days. From that point on, your site would be connected to that other page, and subsequent visitors to your site could follow that connection with a single mouse-click. In some basic sense, by linking to the original restaurant site, you would be recycling the information stored there, making your review more informative. Another food lover might stumble across your review and link to it from her own site, or forward the URL for your review in an e-mail message to a few friends. But for the most part, the information added to the system would remain trapped on your original page, like a lonely cactus waiting for a handful of insects to stumble across it.

Fast-forward to the present. You're sitting in the same restaurant, having just finished a delightful bowl of vichyssoise, and you pull out your mobile phone and compose a 140-character rave review of the soup, with a link to the restaurant's website, and you post it to Twitter before the check has even arrived. Just as before, you are adding new information to the Web's ecosystem with that tweet. But what happens to that information after you press "submit" on your phone?

For starters, it circulates through the ecosystem in a way that was unthinkable in 1995. Within seconds of your composing the note, it is pushed out to all your Twitter followers, in some cases sent directly to their mobile phones. Thanks to the "re-tweeting" convention spontaneously adopted by the Twitter community, that original vichyssoise tweet is easily forwarded along to other foodies on Twitter. But that's just the start of the journey. Thanks to the geographic data attached to the post by your GPS-powered mobile device, the real-world social network Foursquare automatically distributes the vichyssoise tweet to all its users who have recently visited nearby bars, restaurants, or other public spaces. (Even coffeehouses!) The tweet pops up immediately as a pushpin on the countless Twitter-maps that developers have created over the past few years. The hy-perlocal news platform Outside.in (which I helped create a few years ago) parses the geo-data and detects the name of the restaurant in the tweet, and automatically attaches it to pages devoted to discus-sion of the restaurant itself, along with pages that cover all the news and commentary about the Back Bay neighborhood, and pages de-voted to the Boston restaurant scene. A Boston newspaper that has built neighborhood-specific news pages using Outside.in's open pub-lisher platform runs that tweet on a page devoted to food gossip in the Back Bay. Google detects the link to the restaurant's website and registers the link as a "vote" endorsing the quality of that page, which causes it to rise higher in the search-results page when people query Google with its name. The tweet even shows up in the inbox of the restaurant's proprietor, who has established a Google Alert that automatically e-mails him when anything appears online that mentions his restaurant by name. On many of these pages—on the newspaper sites, on Google—local ads appear for other businesses in the neighborhood, drawn like moths to the bright flame of the geo-graphic data embedded in the tweet.

Most of that whole sequence unfolds within minutes, without you having to think about anything other than composing those 140 characters and remembering to press "submit."

The story here is not the old chestnut of living in a connected age where information flows more quickly than ever before. The information is not simply *flowing* in this system; it's being *recycled* and put to new uses, transformed by a diverse network of other species in the ecosystem, each with its own distinct function. You write a tweet about what you had for lunch—the original sin of

Twitter banality—and within minutes that information is being harnessed to assist a staggering number of different tasks: neighbors forging new personal connections, foodies seeking a delicious cup of potato and leek soup, restaurant owners getting unvarnished feedback from their patrons, Google organizing all the world's information, newspapers improving their neighborhood coverage at lower cost, and local businesses seeking the attention of the people in their immediate community. Not bad for 140 characters.

But of course the point is that those 140 characters had help. At every step of their journey, they were standing on layers of stacked platforms. The simplicity of sending out a message to a social network of followers depends on the Twitter API and underlying database; that they instantly reach mobile phones as text messages relies on the SMS communications protocol (along with the network of cell towers and satellites); Outside.in distributes its neighborhood data using the open RSS platform; the geo-data embedded in the original tweet relies on the adapted military intelligence technology of GPS; the Twittermaps all involve API calls to Google's map service; and, of course, the entire operation is sustained by the coral-and-zooxanthella foundation of underlying protocols like HTTP and TCP/IP. All those services and standards were essential to the web of information that benefited from those 140 characters, but not one of them required a business development deal, or a licensing fee, or even an old-fashioned handshake. You can build on all of them without asking for permission, and when you don't have to ask for permission, innovation thrives. When Guier, Weiffenbach, and McClure were designing their system to help American submarines launch Polaris missiles against the Soviet Union, it never occurred to them that someday someone would use their platform to rave about a bowl of potato and leek soup to nearby strangers. Stacked platforms are like that: you think you're fighting the Cold War, and it turns out you're actually helping people figure out where to have lunch.

In a funny way, the real benefit of stacked platforms lies in the knowledge you no longer need to have. You don't need to know how to send signals to satellites or parse geo-data to send that tweet circulating through the Web's ecosystem. Miles Davis didn't have to build a valved trumpet or invent the D Dorian mode to record *Kind of Blue*. The songbird sitting in an abandoned woodpecker's nest doesn't need to know how to drill a hole into the side of a poplar, or how to fell a hundred-foot tree. That is the generative power of open platforms. The songbird doesn't carry the cost of drilling and felling because the knowledge of how to do those things was openly supplied by other species in the chain. She just needs to know how to tweet.

Conclusion

THE FOURTH QUADRANT

On the somewhat desolate corner of Grand Street and Morgan Avenue in the Williamsburg neighborhood of Brooklyn, a five-story building stands, built in the watered-down Romanesque style favored by industrial architects a century ago. Today it is home to a mix of uses: twentysomething roommates sharing loft spaces on the fringes of one of New York's hottest neighborhoods, amid a handful of small businesses, most of them in the information industries. A hundred years ago, the building had a single tenant: the Sackett-Wilhelm Lithography Company. If you stand at the front door on Grand Street, or scan the bars on the first-floor windows and the graffiti on the old loading docks, nothing indicates the historic nature of the site. But historic it is: the Sackett-Wilhelm Lithography Company housed the first working version of a machine that would do more to transform the settlement patterns of human beings than any other twentieth-century invention, with the possible exception of the automobile.

In 1902, the Sackett-Wilhelm company had a profitable and growing business printing color publications, like the popular humor magazine *Judge*. But they faced one vexing problem: the air. Small changes in humidity could complicate the printing process on multiple levels: the paper would expand as it absorbed water molecules floating in the factory air; ink would flow at different rates, and dry more slowly. Unusually humid days could slow the entire production down dramatically, making it difficult for the Sackett-Wilhelm executives to promise reliable delivery times to its clients.

Human beings had been artificially moderating air temperature since the invention of fire. The nineteenth century had witnessed a growing trend toward mechanical heating systems. A few exotic schemes had attempted to cool building interiors, but all of them involved drawing air over massive quantities of ice. (Madison Square Theater in Manhattan used four tons of ice each night to make summertime evenings tolerable for their patrons.) But none of these approaches tackled the problem of humidity. After two consecutive heat waves in the summers of 1900 and 1901, the Sackett-Wilhelm owners contacted the New York office of the Buffalo Forge Company, which specialized in mechanical heating systems for large industry. They were the experts in making the air warmer. Could they make it less wet?

It was a fortuitous query, because the founder of the Buffalo Forge Company, William F. Wendt, had just caved in to the demands of an ambitious twenty-five-year-old electrical engineer named Willis Carrier and created a "research program" where Carrier could take on more speculative projects. Carrier's lab was the perfect place to tackle a problem like dehumidifying air, and Carrier threw himself into the project with enthusiasm. After experimenting with a handful of failed schemes suggested by his colleagues, Carrier followed his own instincts and built a contraption that passed chilled water through a heating coil that usually conveyed steam. Using dew-point charts from the Weather Bureau, he built a system that cooled the air to the dew-point temperature that would produce the 55-percent humidity that the Sackett-Wilhelm company considered optimal. By the late summer of 1902, a system engineered by Carrier was operational in the Sackett-Wilhelm plant. It drew water from an artesian well, with additional cooling provided by an ammonia refrigerating machine. The overall cooling effect on a hot summer day was the equivalent of melting 108,000 pounds of ice in a single twenty-four-hour period.

Carrier would continue tinkering with his system over the following years. The Sackett-Wilhelm system had been a success, but the steel coils were prone to rust after regular use. One night, waiting for a train in Philadelphia, watching a heavy fog roll across the platform, he had a sudden flash of insight. His air-conditioning system could be a miniature fog machine: by drawing air across a fine spray of water inside the device, he could use the water itself as a condensing surface. Thanks to those tenacious hydrogen bonds, the molecules of water vapor in the spray would pull the moisture out of the air, regulating the humidity and eliminating the rust problem. (As Carrier put it in his autobiography: "Water won't rust.") Carrier applied for a patent for his "Apparatus for Treating Air" in September of 1904. On the second day of 1906, the patent was granted. Before long, Carrier and a band of entrepreneurial engineers from Buffalo Forge broke off and formed the Carrier

Engineering Corporation, devoted exclusively to the manufacture of air-conditioning systems. The business made Carrier a wealthy man, as air conditioning went from a curiosity to a luxury item to a middle-class necessity. In 2007, the Carrier Corporation, now part of United Technologies, did $15 billion in sales. Thanks to Carrier's brilliant idea, the second half of the twentieth century saw a mass migration within the United States to the Sunbelt and to Deep South climates that had been nearly intolerable before the widespread adoption of air conditioning. It is not exaggerating matters to say that Carrier's idea ultimately rearranged the social and political map of America.

Carrier's story is the archetypal myth of modern innovation. A clever individual, working in a private research lab, driven by ambition and the promise of great riches, hits upon a brilliant idea in a sudden flash of insight and the world changes. Yes, Carrier's story is slightly more complicated than this cartoon version suggests. He was more focused originally on humidity than on temperature; the ultimate solution took several years to crystallize; and some of his technical solutions built on the ideas of those who had come before him. But this is quibbling. Carrier's narrative fits the classic mold of the genius entrepreneur. It lacks almost all of the patterns that we have seen over the preceding chapters: no liquid networks (if you don't count the fog); no coffeehouse exaptations; no brilliant mistakes. And it ends with a triumphant patent grant.

All of which leads to the inevitable question: Is Willis Carrier an anomaly or not?

The question has real political and social stakes, because the doxa of market capitalism as an unparalleled innovation engine has long leaned on stories like Willis Carrier's miraculous cooling device as a cornerstone of its faith.[6] In many respects, these beliefs made sense, because the implicit alternatives were the planned economies of socialism and communism. State-run economies were fundamentally hierarchies, not networks. They consolidated decision-making power in a top-down command system, which meant that new ideas had to be approved by the authorities before they could begin to spread through the society. Markets, by contrast, allowed good ideas to erupt anywhere in the system. In modern tech-speak, markets allowed innovation to flourish at the edges of the network. Planned economies were more like the old mainframe computer systems that predated the Internet, where every participant had to get authorization from a central machine to do new work. When Friedrich von Hayek launched his influential argument in the 1940s about the importance of price signals in market economies, he was observing a related phenomenon: the decentralized pricing mechanism of the marketplace allows an entrepreneur to gauge the relative value of his or her innovation. If you come up with an interesting new contraption, you don't need to persuade a government commission of its value. You just need to get someone to buy it.

---

6. Innovation, of course, is not the sole reason so much of the world defied the predictions of *The Communist Manifesto* and embraced the capitalist way of life. Economists and social historians have documented multiple factors that drove the march of the market: capitalist economies had a better track record of long-term increases in GDP; economic actors had more liberty to make individual choices; economic self-interest is an undeniable motivating force for human beings. But few defenses of capitalism's economic virtues failed to mention its protean force. Even its critics acknowledged the market's drive for novelty and innovation, as in Joseph Schumpeter's famous theory of "creative destruction."

Entire institutions and legal frameworks—not to mention a vast tower of conventional wisdom—have been built around the Carrier model of innovation. But what if he's the exception and not the rule?

There are three main approaches for settling a question as complicated as this. You can dive deeply into a single story and try to persuade your audience that it is representative of a larger societal truth. (This is the strategy I adopted in telling the stories of John Snow and Joseph Priestley—and the innovation environments that shaped their work—in my previous two books.) The advantage of this approach is that it allows you to examine a case study in exhaustive detail. The disadvantage, of course, is that your audience has to take it on faith that the case study you've chosen is indeed representative of a wider truth. The second approach, which I have taken in the preceding chapters of this book, is to build an argument around dozens of anecdotes, drawn from different contexts and historical periods. The anecdotal approach sacrifices detail for breadth. Yet it, too, runs the risk of being accused of cherry-picking. If there are a hundred Willis Carriers for every Tim Berners-Lee, it doesn't really prove anything to string together a book of Berners-Lee stories. (In fact, it may well be misleading to do so.)

To see around the potential distortions of the case-study and anecdotal approaches, you need to see the entire field of innovation through a single lens. You can't tell whether Willis Carrier is an anomaly by studying the fine points of his biography. You need a wider view. So let us perform an experiment on the data available on the history of innovation. Take roughly two hundred of the most important innovations and scientific breakthroughs from the past six hundred years, starting with Gutenberg's press: everything from

Einstein's theory of relativity to the invention of air conditioning
to the birth of the World Wide Web. Plot each breakthrough somewhere in one of the four quadrants of this diagram:

Classify innovations that involved a small, coordinated team
within an organization—or, even better, a single inventor—as "individual." Classify as "networked" all the innovations that evolved
through collective, distributed processes, with a large number of
groups working on the same problem. Inventors who planned to

capitalize directly from the sale or licensing of their invention
should be classified as "market"; those who wished their ideas to
flow freely into the infosphere belong to the "non-market" side.
The result is four quadrants: the first correlating to the private corporation or the solo entrepreneur; the second to a marketplace
where multiple private firms interact; the third to the amateur scientist or hobbyist who shares his or her ideas freely; and, finally, the
fourth quadrant, which corresponds to open-source or academic environments, where ideas can be built upon and reimagined in large,
collaborative networks.

By taking this long view, we can begin to answer the question
we began with: Just how dominant is the Willis Carrier model of
innovation?[7] Which quadrant has the most impressive track record
for generating good ideas?

To give us some bearings, our anchor tenant in the first
quadrant—the market-based individual—is Carrier himself, who
single-handedly drove the invention of air conditioning and who
had clear commercial aspirations for his device. (Gutenberg belongs
there as well.) An example of a networked market innovation would
be the vacuum tube, the creation of which involved a decentralized
network with dozens of key participants, including Lee de Forest,
almost all of whom worked either as patent-prone entrepreneurs or
research scientists within larger corporations. Tim Berners-Lee's

---

7. This framework is adapted from Yochai Benkler's book *The Wealth of Networks*. Benkler's
point is that we have extensive experience with three of the four possible combinations. Private
corporations are centralized and market-based. The marketplace itself is decentralized and,
obviously, market-based. Planned economies are centralized and non-market-based. But the
magic square is the fourth one: that of decentralized, non-market environments. This is a combination that does not easily fit into the standard boxes of capitalism and socialism. Yet in recent
years, this quadrant has been a hothouse of innovation, thanks in large part to the open architecture of the Internet.

creation of the World Wide Web belongs to the individual, non-market quadrant, while the Internet itself belongs to the fourth quadrant, given the vast number of public sector individuals and organizations involved in its creation.

It should be noted that these classifications do not reflect the cumulative nature of almost any innovation. Berners-Lee needed the open platform of the Internet for his hypertext creation to take flight, and thus the many individuals who built ARPANET and TCP/IP should be understood as essential contributors to the Web. Had those platforms been more proprietary ones—say, by charging licensing fees for the privilege of developing on top of them—it's entirely possible that Berners-Lee wouldn't have bothered creating the Web in the first place, given that it was a side project that his superiors knew next to nothing about.

It is in the nature of good ideas to stand on the shoulders of the giants who came before them, which means that by some measure, every important innovation is fundamentally a network affair. But, for the sake of clarity, let's not blur the line between "individual" and "network" by admitting to the discussion the prior innovations that inspired or supported the new generation of ideas. Yes, it is important that Gutenberg borrowed the screw-press technology from the winemakers, but one cannot say that the printing press was a collective innovation the way, for example, the Internet clearly was. So Gutenberg and Berners-Lee get classified on the individual side of the spectrum.

There is no reliable mathematical formula for making these classifications, and to a certain extent each of them involves an element of subjectivity. But I think that, seen together as a group, they reveal an interesting pattern—interesting enough, I would argue, for us to tolerate a little noise in the data. We are accustomed to looking at certain historical developments—mostly demographic—in this condensed, time-lapse format. We watch the growth of cities, or markets, or national populations unfold in charts where each tick measures a century. There are truths made visible by these time-lapse views that present-tense surveys or individual, narrative histories cannot properly shine light on. (Malthus's *Principles of Population*, which so inspired Darwin and Wallace, offered an early glimpse of that special effect.) But we rarely measure *cultural* changes this way. So much of the history of ideas is like Darwin's work as a naturalist during the long years that preceded the publication of *Origin*: analyzing an individual species, defining its key characteristics, and putting it in the proper box. That's a fine approach for understanding why a specific idea came into being at a particular moment in time. But if you want to wrestle with the question one link farther up the chain—how do good ideas *tend* to come about—you need to take on the problem from a different angle. There's a place for counting barnacles. But sometimes you need to zoom out and take the longer view.

In taking this approach, I am exapting a technique that the literary historian Franco Moretti calls "distant reading." In a series of influential books and essays published over the past decade, Moretti has broken from the traditional English Department approach of "close reading," in which individual literary texts are analyzed in exhaustive detail. It doesn't really matter whether the close reading in question is an old-school tribute to an artist's singular talents or a politicized deconstruction—you can read the text closely to reveal the genius of the author, or his latent homophobia, but in each case you're doing close reading, where every sentence is

British Novelistic Genres, 1740–1900

Genre	
Kailyard school	
New Woman novel	
Imperial gothic	
Naturalist novel	
Decadent novel	
Nursery stories	
Regional novel	
Cockney school	
Utopia	
Invasion literature	
Imperial romances	
School stories	
Children's adventures	
Fantasy	
Sensation novel	
Provincial novel	
Domestic novel	
Religious novel	
Bildungsroman	
Multiplot novel	
Mysteries	
Chartist novel	
Sporting novel	
Industrial novel	
Conversion novel	
Newgate novel	
Nautical tales	
Military novel	
Silver-fork novel	
Romantic farrago	
Historical novel	
Evangelical novel	
Village stories	
National tale	
Anti-Jacobin novel	
Gothic novel	
Jacobin novel	
Ramble novel	
Spy novel	
Sentimental novel	
Epistolary novel	
Oriental tale	
Picaresque	
Courtship novel	

1700   1750   1800   1850   1900

a potential datapoint in your analysis. ("At bottom," Moretti writes, "it's a theological exercise—very solemn treatment of very few texts taken very seriously.") Distant reading takes the satellite view of the literary landscape, looking for larger patterns in the history of the stories we tell each other. In one typically inventive analysis, Moretti tracked the evolution of subgenres in popular British novels from 1740 to 1915, an immense taxonomy of narrative forms—spy novels, picaresques, gothic novels, nautical tales, mysteries, and dozens of other distinct forms. He plotted the life span of each subgenre as a dominant species in the British literary ecosystem. The result is on page 223.

What happens when you take the distant approaching to reading novels is that you're able to see patterns that simply aren't visible on the scale of paragraphs and pages, or even entire books. You could read a dozen "silver fork" novels and bildungsromans and yet miss the most striking fact revealed by Moretti's chart: that the diversity of forms is strikingly balanced by their uncannily similar life spans, which Moretti attributes to underlying generational turnover. Every twenty-five to thirty years a new batch of genres becomes dominant, as a new generation of readers seeks out new literary conventions. If you're trying to understand the meaning of an individual work, you have to read closely. But if you're interested in the overall behavior of the literary system—its own patterns of innovation—sometimes you have to read from a long way off.

In the study of scientific or technological innovation, the equivalent of close reading is the meticulous biography of the great inventor, or the history of a single technology: the radio, say, or the personal computer. As valuable as those approaches can be, they have their limitations. Close reading leaves you with the idiosyncrasies of

each individual or invention, the local color—but not the general
laws. When you view the history of innovation from a distance, what

1
Printing Press
Mercator Projection

2
Portable Watches
Double-Entry Bookkeeping
Stocking Frame

MARKET/INDIVIDUAL   MARKET/NETWORK

NON-MARKET/INDIVIDUAL   NON-MARKET/NETWORK

3
Concave Lens   Pulmonary Circulation
Terrestrial Globe   Supernovas
Earth Rotates Around Sun   Comet
Steam Turbine   Flush Toilet
Parachute
Ball Bearings
Square Root and Plus and
Minus Symbols
Ether

4
Pencil
Microscope
Cubic Equations/Complex
Numbers

1400–1600

you lose in detail you gain in perspective. Classifying two hundred good ideas into four broad quadrants certainly makes it harder to learn anything specific about each individual innovation. But it does allow us to answer the question we began with: What kind of environments make innovation possible in the first place?

Because innovation is subject to historical changes—many of which are themselves the result of influential innovations in the transmission of information—the four quadrants display distinct shapes at different historical periods. Start with this view of the breakthrough ideas from 1400 to 1600, beginning with Gutenberg's press and continuing on to the dawn of the Enlightenment (see page 225).

This is the shape that Renaissance innovation takes, seen from a great (conceptual) distance. Most innovation clusters in the third quadrant: non-market individuals. A handful of outliers are scattered fairly evenly across the other three quadrants. This is the pattern that forms when information networks are slow and unreliable, and entrepreneurial economic conventions are poorly developed. It's too hard to share ideas when the printing press and the postal system are still novelties, and there's not enough incentive to commercialize those ideas without a robust marketplace of buyers and investors. And so the era is dominated by solo artists: amateur investigators, usually well-to-do, working on their own private obsessions. Not surprisingly, this period marks the birth of the modern notion of the inventive genius, the rogue visionary who somehow sees beyond the horizon that limits his contemporaries—da Vinci, Copernicus, Galileo. Some of those solo artists (Galileo most famously) worked outside of broader groups because their research posed a significant security threat to the established powers of the day. The few innovations that did emerge out of networks—the

1

Pressure Cooker
Manned Hot Air Balloon
Lithography

2

Chronometer
Balance Spring Watches
Steam Engine
Steamboat
Spinning Jenny
Power Loom
Cotton Gin

MARKET/INDIVIDUAL | **MARKET/NETWORKED**

NON-MARKET/INDIVIDUAL | **NON-MARKET/NETWORKED**

3

Magnetism (Planetary)	Moons of Jupiter
Speed of Light	Microorganisms
Piano	Laws of Motion
Tuning Fork	Comet Orbits
Mercury Thermometer	Lightning Rods
Octant	Bifocal Lenses
Carbonated Water	Plant Respiration
Hooke's Law	Logarithms
Law of Falling Bodies	Blood Circulation
Elliptical Orbits of Planets	Vernier Scale
	Analytic Geometry

4

Vacuum Pump	Photosynthesis
Pendulum Clock	Milky Way
Boyle's Law	Hot Air Balloon
Light Spectrum	Smallpox Vaccine
Calculus	Sunspots
Universal Gravitation	Slide Rule
Flintlock	Mechanical Calculator
Telescope	Ocean Tides
Linnaean Taxonomy	Barometer
Discovery of Oxygen	

1600–1800

portable, spring-loaded watches that first appeared in Nuremberg in 1480, the double-entry bookkeeping system developed by Italian merchants—have their geographic origins in cities, where information networks were more robust. First-quadrant solo entrepreneurs, crafting their products in secret to ensure their eventual payday, turn out to be practically nonexistent. Gutenberg was the exception, not the rule.

Scanning the next two centuries, we see that the pattern changes dramatically (see page 227).

Solo, amateur innovation (quadrant three) surrenders much of its lead to the rising power of networks and commerce (quadrant four). The most dramatic change lies along the horizontal axis, in a mass migration from individual breakthroughs (on the left) to the creative insights of the group (on the right). Less than 10 percent of innovation during the Renaissance is networked; two centuries later, a majority of breakthrough ideas emerge in collaborative environments. Multiple developments precipitate this shift, starting with Gutenberg's press, which begins to have a material impact on secular research a century and a half after the first Bible hits the stands, as scientific ideas are stored and shared in the form of books and pamphlets. Postal systems, so central to Enlightenment science, flower across Europe; population densities increase in the urban centers; coffeehouses and formal institutions like the Royal Society create new hubs for intellectual collaboration.

Many of those innovation hubs exist outside the marketplace. The great minds of the period—Newton, Franklin, Priestley, Hooke, Jefferson, Locke, Lavoisier, Linnaeas—had little hope of financial reward for their ideas, and did everything in their power to encourage their circulation. A vertical movement toward market incentives

MARKET/INDIVIDUAL

Mason Jar
Tesla Coil
Gatling Gun
Nylon
Vulcanized Rubber
Programmable Computer
Revolver
Dynamite
AC Motor
Air-Conditioning
Transistor

MARKET/NETWORKED

Airplane	Lightbulb
Steel	Automobile
Induction Motor	Radio
Contact Lenses	Welding Machine
Moving Assembly Line	Motion Picture Camera
Locomotive	Vacuum Cleaner
Electric Motor	Washing Machine
Refrigerator	Vacuum Tube
Telegraph	Helicopter
Sewing Machine	Television
Elevator	Photography
Steel	Jet Engine
Typewriter	Tape Recorder
Plastic	Laser
Calculator	VCR
Internal Combustion	Personal Computer
Engine	Bicycle
Telephone	

NON-MARKET/INDIVIDUAL

Spectroscope	Hormones
Bunsen Burner	$E = mc^2$
Rechargeable Battery	Special Relativity
Nitroglycerine	Earth's Core
Liquid Engine Rocket	Radiometric Dating
Uncertainty Principle	Cosmic Radiation
Electrons in Chemical	General Relativity
Bonds	Universe Expanding
Absolute Zero	Ecosystem
Atomic Theory	Double Helix
Stethoscope	CT Scan
Uniformitarianism	Archaea
Cell Nucleus	World Wide Web
Benzene Structure	Continental Drift
Heredity	Superconductors
Natural Selection	Neutron
X-Rays	Early Life Simulated
Blood Groups	

NON-MARKET/NETWORKED

Braille   Periodic Table   RNA Splicing
Chloroform   EKG   Cosmic Microwave Background Radiation
Aspirin   Cell Division   Global Warming   MRI
Enzymes   Cell Differentiation   DNA Forensics
Stratosphere   Radioactivity   Plate Techtonics
Cosmic Rays   Electron   Atomic Reactor
Modern Computer   Mitochondria   Nuclear Forces
Artificial Pacemaker   Vitamins   Oral Contraceptive
Radiocarbon Dating   Neurotransmitters
Graphic Interface   Genes on Chromosomes
Endorphins   Chemical Bonds   Restriction Enzymes
Infant Incubator   Radiography   Gamma-Ray Bursts
Oncogenes   Penicillin   Universe Accelerating
Atoms Form Molecules   Quantum Mechanics
Punch Cards (Jacquard Loom)   Radar   GPS
Suspension Bridge   Liquid-Fueled Rocket
Second Law   DNA (as Genetic Material)   Internet
Anesthesia   Krebs Cycle   RNA (as Genetic Material)
Germ Theory   Computer   Asteroid K-T Extinction

1800–present

is noticeable, nonetheless. As industrial capitalism arises in England
in the eighteenth century, new economic structures raise the stakes
for commercial ventures: tantalizing rewards lure innovators into
private enterprise, and the codification of English patent laws in the
early 1700s gives some reassurance that good ideas will not be stolen
with impunity. Despite this new protection, most commercial innovation during this period takes a collaborative form, with many
individuals and firms contributing crucial tweaks and refinements
to the product. The history books like to condense these slower,
evolutionary processes into eureka moments dominated by a single
inventor, but most of the key technologies that powered the Industrial Revolution were instances of what scholars call "collective invention." Textbooks casually refer to James Watt as the inventor of
the steam engine, but in truth Watt was one of dozens of innovators
who refined the device over the course of the eighteenth century.

Let us pause for a moment on the cusp of the modern age and
take a few bets as to what pattern will form in the final two
centuries of the millennium. I think most of us would expect to see
a dramatic consolidation of innovative activity in the first quadrant,
as capitalism enters its mature period, spanning the ages of mass
production and the consumer society. All the elements would seem
to predict an explosion of first-quadrant activity: an increasingly
wealthy public willing to spend money on new gadgets; strong enforcement of intellectual property rights; the emergence of corporate research-and-development labs; and a growing pool of private
capital willing to finance speculative ventures. If the competitive
marketplace of modern capitalism is the great innovation engine of

our time, the first quadrant should by rights dominate the last two centuries of activity.

But instead, another pattern appears (see page 229).

Against all odds, the first quadrant turns out to be the least populated on the grid. Willis Carrier is an outlier after all. In the private sector, the proprietary breakthrough achieved in a closed lab turns out to be a rarity. For every Alfred Nobel, inventing dynamite in secret in the suburbs of Stockholm, there are a half dozen collective inventions like the vacuum tube or the television, whose existence depended upon multiple firms driven by the profit motive who managed to create a significant new product via decentralized networking. Folklore calls Edison the inventor of the lightbulb, but in truth the lightbulb came into being through a complex network of interaction between Edison and his rivals, each contributing key pieces to the puzzle along the way. Collective invention is not some socialist fantasy; entrepreneurs like Edison and de Forest were very much motivated by the possibility of financial rewards, and they tried to patent as much as they could. But the utility of building on other people's ideas often outweighed the exclusivity of building something entirely from scratch. You could develop small ideas in a locked room, cut off from the hunches and insights of your competition. But if you wanted to make a major new incursion into the adjacent possible, you needed company.

Even more striking, though, is the explosion of fourth-quadrant activity.

Why have so many good ideas flourished in the fourth quadrant, despite the lack of economic incentives? One answer is that economic incentives have a much more complicated relationship to the development and adoption of good ideas than we usually imag-

ine. The promise of an immense payday encourages people to come
up with useful innovations, but at the same time it forces people to
protect those innovations. Economists define "efficient markets" as
markets where information is evenly distributed among all the buyers and sellers in the space. Efficiency is generally held to be a
universal goal for any economy—unless the economy happens to
traffic in ideas. If ideas were fully liberated, then entrepreneurs
wouldn't be able to profit from their innovations, because their competitors would immediately adopt them. And so where innovation
is concerned, we have deliberately built *inefficient* markets: environments that protect copyrights and patents and trade secrets and
a thousand other barricades we've erected to keep promising ideas
out of the minds of others.

That deliberate inefficiency doesn't exist in the fourth quadrant. No, these non-market, decentralized environments do not have
immense paydays to motivate their participants. But their openness
creates other, powerful opportunities for good ideas to flourish. All
of the patterns of innovation we have observed in the previous
chapters—liquid networks, slow hunches, serendipity, noise, exaptation, emergent platforms—do best in open environments where
ideas flow in unregulated channels. In more controlled environments, where the natural movement of ideas is tightly restrained,
they suffocate. A slow hunch can't readily find its way to another
hunch that might complete it if there's a tariff to be paid every
time it tries to make a new serendipitous connection; exaptations
can't readily occur across disciplinary lines if there are sentries
guarding those borders. In open environments, however, those patterns of innovation can easily take hold and multiply.

Like any complex social reality, creating innovation environments is a matter of trade-offs. All other things being equal, financial incentives will indeed spur innovation. The problem is, all other things are *never* equal. When you introduce financial rewards into a system, barricades and secrecy emerge, making it harder for the open patterns of innovation to work their magic. So the question is: What is the right balance? It's certainly conceivable that the promise of hitting a financial jackpot is so overwhelming that it more than makes up for the inefficiencies introduced by intellectual property law and closed R&D labs. That has generally been the guiding assumption for most modern discussions of innovation's roots, an assumption largely based on the free market's track record for innovation during that period. Because capitalist economies proved to be more innovative than socialist and communist economies, the story went, the deliberate inefficiencies of the market-based approach must have benefits that exceed their costs.

But, as we have seen, this is a false comparison. The test is not how the market fares against command economies. The real test is how it fares against the fourth quadrant. As the private corporation evolved over the past two centuries, a mirror image of it grew in parallel in the public sector: the modern research university. Most academic research today is fourth-quadrant in its approach: new ideas are published with the deliberate goal of allowing other participants to refine and build upon them, with no restrictions on their circulation beyond proper acknowledgment of their origin. It is not pure anarchy, to be sure. You can't simply steal a colleague's idea without proper citation, but there is a fundamental difference between suing for patent infringement and asking for a footnote. Academics are paid salaries, of course, and successful ideas can lead to much-sought-after tenured professorships, but the economic rewards are minuscule compared to those of the private sector. And, crucially, those rewards are not dependent on introducing an artificial inefficiency into the information network. A historian who develops a brilliant new theory about the origins of the Industrial Revolution may well land a chaired professorship at an Ivy League school thanks to her theory, but the theory itself can freely circulate through the environment, where it can be challenged, enlarged, exapted, and recycled in countless ways. The university system may be big business these days, and patents do play a role in some specialized fields, but for the most part the university remains an information commons.

Universities have a reputation for ivory-tower isolation from the real world, but it is an undeniable fact that most of the paradigmatic ideas in science and technology that arose during the past century have roots in academic research. This is obviously true for the "pure" sciences like theoretical physics, but it is also true for lines of research that on their surface seem to have more straightforwardly commercial applications. The oral contraceptive, for instance, has generated billions of dollars for Big Pharma over the past half century, but most of the critical research that led to its development happened in the intellectual commons of university labs at Harvard, Princeton, and Stanford. In the language of the last chapter, open networks of academic researchers often create emergent platforms where commercial development becomes possible. The next decade will likely see a wave of pharmaceutical products enabled by genomic science, but that underlying scientific platform—most critically, the ability to sequence and map DNA— was almost entirely developed by a decentralized group of academic scientists working outside the private sector in the 1960s and seventies. This is a pattern we see again and again in the modern era: fourth-quadrant innovation creates a new open platform that commercial entities can then build upon, either by repackaging and refining the original breakthrough, or by developing emergent innovations on top of the underlying platform.

Fourth-quadrant innovation has been assisted by another crucial development: the increased flow of information. Information spillover required the geographic density of cities in the Renaissance, while the postal system made small distributed webs of creativity possible in the Enlightenment. But the Internet has effectively reduced the transmission costs of sharing good ideas to zero. In Galileo's time, all the benefits of information spillover were as potent as they are today. But it was far more difficult to create the kind of liquid network where those serendipitous collisions and exaptations could take place. The connectedness of modern life means that we face the opposite problem: it is much harder to *stop* information from spilling over than it is to get it into circulation. The consequence of this is that private-sector firms who are intent on protecting their intellectual assets have to invest time and money in building barricades of artificial scarcity. Participants in the fourth quadrant don't have those costs: they can concentrate on coming up with new ideas, not building fortresses around the old ones. And because those ideas can freely circulate through the infosphere, they can be refined and expanded by other minds in the network.

We do not have a ready-made political vocabulary for the fourth quadrant, particularly the noninstitutional forms of collaboration that have developed around the open-source community. Because these open systems operate outside the conventional incentives of capitalism and resist the usual strictures of intellectual property, the mind reflexively wants to put them on the side of socialism. And yet they are as far from the state-centralized economies that Marx and Engels helped invent as they are from greed-is-good capitalism. They themselves are not the product of market incentives, but they often create environments where private firms thrive, a phenomenon Lawrence Lessig alludes to in his concept of the "hybrid economy," which blends elements from the open networks of the intellectual commons with the more proprietary walls and tariffs of the private sphere.

None of this is meant to imply that the marketplace is the enemy of innovation, or that competition between rival firms doesn't often lead to useful new products. (The second quadrant, after all, bustles with dozens of brilliant ideas that changed our lives for the better.) And top-heavy bureaucracies remain innovation sinkholes. But, fortunately for us, the choice is not between decentralized markets and command-and-control states. Much of the history of intellectual achievement over these past centuries has lived in a less formal space between those two regimes: in the grad seminar and the coffeehouse and the hobbyist's home lab and the digital bulletin board. The fourth quadrant should be a reminder that more than one formula exists for innovation. The wonders of modern life did not emerge exclusively from the proprietary clash between private firms. They also emerged from open networks.

A few months after Darwin published *On the Origin of Species* in 1859, Karl Marx wrote Friedrich Engels a letter that included a few lines endorsing Darwin's biological radicalism. "Although it is developed in the crude English style, this is the book which contains the basis in natural history for our view." The "crude English style" was evidently Darwin's strange unwillingness to incessantly relate his scientific views back to Hegelian dialectics. (Many now regard that as one of Darwin's strengths as a writer.) Beyond the sneers, Marx and Engels were clearly energized by the controversy Darwin had unleashed and saw him as a kindred spirit in an age that seemed on the verge of multiple revolutions—in science as well as in society. It is unclear whether Darwin felt quite the same way about his Prussian admirers. Marx offered to dedicate volume two of *Das Kapital* to Darwin, who demurred: "I should prefer the part or volume not to be dedicated to me (although I thank you for the intended honour), as that would, in a certain extent, suggest my approval of the whole work, with which I am not acquainted."

From a scientific point of view, Marx and Engels were smart to side with Darwin so early in the debate over his "dangerous" idea. But they couldn't have been more wrong in their predictions about the way the theory would play out in the politico-economic arena. They anticipated, correctly, that analogies would be drawn between Darwin's "survival of the fittest" and the competitive selection of capitalist free-market economies. Marx and Engels just assumed those analogies would be launched as *critiques* of capitalism. In 1865, Engels wrote to a friend, "Nothing discredits modern bourgeois development so much as the fact that it has not yet succeeded in getting beyond the economic forms of the animal world."

As it turned out, the exact opposite happened. Darwin's theories were invoked countless times in the twentieth century as a defense of the free-market system. Aligning them with the animal world didn't discredit markets, as Engels had predicted. It made markets look *natural.* If Mother Nature made such a splendidly diverse planet through an algorithm of ruthless competition between selfish agents, why shouldn't our economic systems follow the same rules?

Yet the true story of nature is not one of exclusively ruthless competition between selfish agents, as Darwin himself realized. *Origin of Species* ends with one of the most famous passages in the history of science, one that echoes the journal entry he wrote on leaving the Keeling Islands more than twenty years before:

> It is interesting to contemplate a tangled bank, clothed with many plants of many kinds, with birds singing on the bushes, with various insects flitting about, and with worms crawling through the damp earth, and to reflect that these elaborately constructed forms, so different from each other, and dependent upon each other in so complex a manner, have all been produced by laws acting around us . . . Thus, from the war of nature, from famine and death, the most exalted object which we are capable of conceiving, namely, the production of the higher animals, directly follows. There is grandeur in this view of life . . .

Darwin's words here oscillate between two structuring metaphors that govern all his work: the complex interdependencies of the tangled bank, and the war of nature; the symbiotic connections of an ecosystem and the survival of the fittest. The popular caricature of Darwin's theory emphasizes competitive struggle above everything else. Yet so many of the insights his theory made possible have revealed the collaborative and connective forces at work in the natural world.

We have been living with a comparable caricature in our assumptions about cultural innovation. Look at the past five centuries from the long view, and one fact confronts the eye immediately: market-based competition has no monopoly on innovation. Competition and the profit motive do indeed motivate us to turn good ideas into shipping products, but more often than not, the ideas themselves come from somewhere else. Whatever its politics, the fourth quadrant has been an extraordinary space of human creativity and insight. Even without the economic rewards of artificial scarcity, fourth-quadrant environments have played an immensely important role in the nurturing and circulation of good ideas—now more than ever. In Darwin's language, the open connections of the tangled bank have been just as generative as the war of nature. Stephen Jay Gould makes this point powerfully in the allegory of his sandal collection: "The wedge of competition has been, ever since Darwin, the canonical argument for progress in normal times," he writes. "But I will claim that the wheel of quirky and unpredictable functional shift (the tires-to-sandals principle) is the major source of what we call progress at all scales." The Nairobi entrepreneur selling sandals in an open-air market may indeed be in competition with other cobblers, but what makes his trade possible is the junkyard full of tires waiting to be freely converted into footwear, and the fact that the good idea of converting tires into sandals can be passed from cobbler to cobbler by simple observation, with no licensing agreements to restrict the flow.

. . .

In 1813, a Boston mill owner, Isaac McPherson, found himself immersed in a long and frustrating patent dispute with a Philadelphia-based inventor named Oliver Evans, who had patented an automated grist mill several years before. Evans's engineering talent was matched only by his litigiousness. He was notorious for aggressively enforcing his patents, and was among the first to exploit the new restrictive powers of the federal patent system after its creation in 1790. The originality of Evans's patented invention was highly debatable; the grist-mill system relied on bucket elevators, conveyor belts, and Archimedean screws—all of which were clearly innovations that had long been in the public domain. When Evans sued McPherson for violating his patents, the Boston industrialist decided to reach out to the first patent commissioner of the United States, a former politician and inventor himself, now living in rural Virginia. And so, in the summer of 1813, McPherson wrote a letter to Thomas Jefferson, asking for his interpretation of Oliver Evans's claim.

Jefferson wrote back on August 13. Reading his letter now, one cannot help but be amazed by the range of Jefferson's intelligence. His focus narrows into intense technical detail on the specifics of Evans's invention, and then widens to their ancient prehistory. ("The screw of Archimedes is as ancient, at least, as the age of that mathematician, who died more than 2,000 years ago. Diodorus Siculus speaks of it, L. i., p. 21, and L. v., p. 217, of Stevens' edition of 1559, folio; and Vitruvius, xii.") He reviews the relevant law with the sharp eye of a legal scholar, opining on the sections that he thinks are fundamentally flawed. But the most stirring passages arise when Jefferson waxes philosophical on the nature of ideas themselves:

> Stable ownership is the gift of social law, and is given late in the progress of society. It would be curious then, if an idea, the fugitive fermentation of an individual brain, could, of natural right, be claimed in exclusive and stable property. If nature has made any one thing less susceptible than all others of exclusive property, it is the action of the thinking power called an idea, which an individual may exclusively possess as long as he keeps it to himself; but the moment it is divulged, it forces itself into the possession of every one, and the receiver cannot dispossess himself of it. Its peculiar character, too, is that no one possesses the less, because every other possesses the whole of it. He who receives an idea from me, receives instruction himself without lessening mine; as he who lights his taper at mine, receives light without darkening me. That ideas should freely spread from one to another over the globe, for the moral and mutual instruction of man, and improvement of his condition, seems to have been peculiarly and benevolently designed by nature, when she made them, like fire, expansible over all space, without lessening their density in any point, and like the air in which we breathe, move, and have our physical being, incapable of confinement or exclusive appropriation. Inventions then cannot, in nature, be a subject of property.

Ideas, Jefferson argues, have an almost gravitational attraction toward the fourth quadrant. The natural state of ideas is flow and spillover and connection. It is society that keeps them in chains.

Does this mean we have to do away with intellectual property law? Of course not. The innovation track record of the fourth quadrant doesn't mean that patents should be abolished and all forms of information allowed to run free. But it should definitely put the lie to the reigning orthodoxy that without the artificial scarcity of intellectual property, innovation would grind to a halt. There are plenty of understandable reasons why the law should make it easier for innovative people or organizations to profit from their creations. We may very well decide as a society that people simply *deserve* to profit from their good ideas, and so we have to introduce a little artificial scarcity to ensure those rewards. As someone who creates intellectual property for a living, I am more than sympathetic toward that argument. But it is another matter altogether to argue that those restrictions will themselves promote innovation in the long run.

As Lawrence Lessig has so persuasively argued over the years, there is nothing "natural" about the artificial scarcity of intellectual property law. Those laws are deliberate interventions crafted by human intelligence and are enforced almost entirely by non-market powers. Jefferson's point, in his letter to McPherson, is that if you really want to get into a debate about which system is more "natural," then the free flow of ideas is always going to trump the artificial scarcity of patents. Ideas are intrinsically copyable in the way that food and fuel are not. You have to build dams to keep ideas from flowing.

To my mind, the great question for our time is whether large organizations—public *and* private, governments *and* corporations alike—can better harness the innovation turbine of fourth-quadrant systems. On the private-sector side, the success of companies like Google and Twitter and Amazon—all of whom have, in different ways, contributed to and benefited from fourth-quadrant innovation—has made it clear that, in the software world, at least, a little openness goes a long way. I suspect those lessons will grow increasingly inescapable in the decades to come. But it is the public sector that I find more interesting, because governments and other non-market institutions have long suffered from the innovation malaise of top-heavy bureaucracies. Today, these institutions have an opportunity to fundamentally alter the way they cultivate and promote good ideas. The more the government thinks of itself as an open platform instead of a centralized bureaucracy, the better it will be for all of us, citizens and activists and entrepreneurs alike.

The wonderful irony is that this historic opportunity comes to governments in part because of an innovation that they unleashed on the world: the Internet, probably the clearest example of the way that public- and private-sector innovation can complement each other. The generative platform of the Internet (and the Web) has created a space where countless fortunes have been made over the past thirty years, but the platform itself was created by the loose affiliation of information scientists around the world, funded, in large part, by the federal government of the United States. There are good ideas, and then there are good ideas that make it easier to have other good ideas. YouTube was a good idea that was made possible by the even better ideas of the Internet and the Web. The fact that those idea-generating platforms were developed outside the private sector is no accident. Proprietary platforms that reach critical mass are not unheard of—Microsoft Windows has had a good run, for instance, and Apple's iPhone platform has been extraordinarily innovative in its first three years—but they are rarities. Generative platforms require all the patterns of innovation we have seen over the preceding pages; they need to create a space where hunches and serendipitous collisions and exaptations and recycling can thrive. It is possible to create such a space in a walled garden. But you are far better off situating your platform in a commons.

B ut perhaps "commons" is the wrong word for the environment we're trying to imagine, though it has a long and sanctified history in intellectual property law. The problem with the term is twofold. For starters, it has conventionally been used in opposition to the competitive struggle of the marketplace. The original "commons" of rural England disappeared when they were swallowed up by the private enclosures of agrarian capitalism in the seventeenth and eighteenth centuries. Yet the innovation environments we have explored are not necessarily hostile to competition and profit. More important, however, the commons metaphor doesn't suggest the patterns of recycling and exaptation and recombination that define so many innovation spaces. When you think of a commons, you think of a cleared field dominated by a single resource for grazing. You don't think of an ecosystem. The commons is a monocrop grassland, not a tangled bank.

I prefer another metaphor drawn from nature: the reef. You need only survey a coral reef (or a rain forest) for a few minutes to see that competition for resources abounds in this space, as Darwin rightly observed. But that is not the source of its marvelous biodiversity. The struggle for existence is universal in nature. The few residents of a desert ecosystem are every bit as competitive as their equivalents on a coral reef. What makes the reef so inventive is not the struggle between the organisms but the way they have learned to collaborate—the coral and the zooxanthellae and the parrotfish borrowing and reinventing each other's work. This is the ultimate explanation of Darwin's Paradox: the reef has unlocked so many doors of the adjacent possible because of the way it shares.

The reef helps us understand the other riddles we began with: the runaway innovation of cities, and of the Web. They, too, are environments that compulsively connect and remix that most valuable of resources: information. Like the Web, the city is a platform that often makes private commerce possible but which is itself outside the marketplace. You do business in the big city, but the city itself belongs to everyone. ("City air is free air," as the old saying goes.) Ideas collide, emerge, recombine; new enterprises find homes in the shells abandoned by earlier hosts; informal hubs allow different disciplines to borrow from one another. These are the spaces that have long supported innovation, from those first Mesopotamian settlements eight thousand years ago to the invisible layers of software that support today's Web.

Ideas rise in crowds, as Poincaré said. They rise in liquid networks where connection is valued more than protection. So if we want to build environments that generate good ideas—whether those environments are in schools or corporations or governments or our own personal lives—we need to keep that history in mind, and not fall back on the easy assumptions that competitive markets are the only reliable source of good ideas. Yes, the market has been a great engine of innovation. But so has the reef.

Most of us, I realize, don't have a direct say in what macro forms of information and economic organization prevail in the wider society, though we do influence that outcome indirectly, in the basic act of choosing between employment in the private or the public sector. But this is the beauty of the long-zoom perspective: the patterns recur at other scales. You may not be able to turn your government into a coral reef, but you can create comparable environments on the scale of everyday life: in the workplaces you inhabit; in the way you consume media; in the way you augment your memory. The patterns are simple, but followed together, they make for a whole that is wiser than the sum of its parts. Go for a walk; cultivate hunches; write everything down, but keep your folders messy; embrace serendipity; make generative mistakes; take on multiple hobbies; frequent coffeehouses and other liquid networks; follow the links; let others build on your ideas; borrow, recycle, reinvent. Build a tangled bank.

Acknowledgments

G iven that this is a book in part about the generative power of slow hunches, it should come as no surprise that the topic has been lingering in my mind for almost a decade now, ever since I designed an elaborate experiment, for my book *Mind Wide Open*, to scan my brain as it was trying to come up with a good idea in an FMRI machine. For the past four years, after I began work on the project in earnest, I have consciously thought of this book as the closing volume in an unofficial trilogy that began with *The Ghost Map* and *The Invention of Air*, both books about world-changing ideas and the environments that made them possible. (In a sense, you can think of this book as the latent theory lurking behind those more focused narrative case-studies.) So I am grateful to the many provocative and serendipitous responses I received from readers, critics, and listeners regarding those two earlier books, many of which opened doors to new rooms that I have greatly enjoyed exploring.

I want to extend particular thanks to several organizations that

supported me during the writing of the book, starting with my colleagues at outside.in, led by Mark Josephson, who have tolerated the
eccentric schedule of an author/executive chairman with gracious-
ness and true friendship. Thanks to Columbia University's Journalism School for appointing me the Hearst New Media Scholar in
Residence and for providing a forum where I could talk about commonplace books, my grad school days, and the iPad in a single lecture. Thanks to the wonderful SXSW festival for inviting me to talk
about the ecosystem of news in the spring of 2009, as the ideas for
this book were starting to come together. My editors at *Time*, *Wired*,
The Wall Street Journal, and *The New York Times*—particularly
Rick Stengel, Alex Star, James Ryerson, Tim O'Brien, Chris Anderson, and Larry Rout—allowed me to work through some of these
ideas (and sentences) in public, and provided insightful comments
along the way. (My former editors at *Discover*, Stephen Petranek
and David Grogan, helped me cultivate some of these themes several years ago during my tenure as a columnist there.)

   As usual, the team at Riverhead has been a great help to me,
both in believing in this idea in its embryonic state, and in allowing me to follow my serendipitous discoveries of both outside.in
and Joseph Priestley. Sean McDonald and Geoff Kloske were incredibly patient when my hunch turned out to be even slower
than originally forecast, and when the book finally began to come
together, they did amazing work turning it into a finished product. I'm also grateful to Matthew Venzon, Emily Bell, Hal Fes-
senden, Helen Conford, and my lecture agents at the Leigh Bureau
for their support. My research assistant, Chris Ross, was enor-
mously helpful in collaborating on our maps of innovation history.
Once again my agent, Lydia Wills, displayed her extraordinary

knack for encouraging my good ideas, and delicately shooting down the idiotic ones.

I am particularly grateful to the people who read parts or all of the manuscript in draft form: Brent Constantz, Charlane Nemeth, Brian Eno, John Wilbanks, and in particular Ray Ozzie, Carl Zimmer, and Scott Berkun, and my favorite line editor, Alexa Robinson. They offered many improvements to the ideas contained in this book. What errors remain are entirely my responsibility. It's up to you to decide whether they prove to be generative ones.

Brooklyn, NY

May 2010

Appendix: Chronology of Key Innovations, 1400–2000

DOUBLE-ENTRY ACCOUNTING (1300–1400)

First codified by the Franciscan friar and mathematician Luca Pacioli in 1494, the double-entry method had been used for at least two centuries by Italian bankers and merchants. Some evidence suggests that the technique was developed by Islamic entrepreneurs who passed it on to the Italians through the trade hubs of Venice and Genoa.

PRINTING PRESS (1440)

While elements of the printing press, including the concept of movable type, date back to earlier Chinese and Korean inventors, the first true printing press that combined the screw press and metallic movable type was created by Johannes Gutenberg circa 1440.

CONCAVE LENS (1451)

Humans have used lenses to magnify images and to start fires for thousands of years, but the first use of a concave lens to treat myopia is attributed to the polymath German cardinal Nicholas of Cusa.

PARACHUTE (1483)

Leonardo da Vinci sketched the original design for a parachute in 1438 in the margin of a notebook. The first physical test of the design occurred in 1783, when Louis-Sébastien Lenormand leapt from the Montpelier Observatory in France and, with the aid of his primitive parachute, landed without injury. In 2000, an exact replica of da Vinci's parachute was constructed and tested, and proved to function.

TERRESTRIAL GLOBE (1492)

The Nuremberg-based mapmaker Martin Behaim constructed the first terrestrial globe in the early 1490s, after returning from extensive journeys in West Africa. He called it the Erdapfel, which translates to "earth apple."

BALL BEARINGS (1497)

Conceived and sketched by Leonardo da Vinci in 1497 as a method to reduce friction; the first patent for ball bearings was awarded to Philip Vaughan in 1794.

PORTABLE WATCHES (1500)

One of the canonical examples of collective invention, portable watches evolved out a group of clockmakers in Nuremberg in the early 1500s, led by one Peter Henlein, who created the first lightweight watch. Heinlen's watch was portable, but not terribly accurate; subsequent improvements by his Nuremberg peers allowed the device to keep better time.

EARTH ROTATES AROUND SUN (1514)

Nicolaus Copernicus first wrote out his "heliocentric" theory of the solar system as a small pamphlet around 1514, but did not formally publish the idea for more than twenty years, for fear of the controversy it would un-leash. Word of his radical theory leaked out and began spreading through the enlightened minds of Europe during that period, but the first official publication came in his posthumous text, *On the Revolutions of the Heavenly Spheres*, published in 1543.

SQUARE ROOT AND PLUS AND MINUS SYMBOLS (1525)

German mathematician Christoph Rudolff invented the modern mathematical symbols "+" and "−" and "√" in *Coss*, the first comprehensive guide to algebra in German in 1525.

CUBIC EQUATIONS AND COMPLEX NUMBERS (1530–1540)

The mathematicians of the Islamic Renaissance published several important papers on the understanding of cubic equations—along with the notion of complex numbers—which are essential to determining the area and volume of objects. But the modern technique for solving them is most prominently associated with the Italian mathematician and engineer Niccolò Tartaglia, who won a famous contest in 1530 that showcased his approach. Two other Italians from that period, Scipione del Ferro and his student Antonio Fiore, contributed to the math as well.

PULMONARY RESPIRATION (1535)

The Spanish religious radical Michael Servetus made the first convincing case that the aeration of blood took place in the lungs, after studying the size of the pulmonary artery as a medical student at the University of Paris.

ETHER (1540)

German botanist Valerius Cordus discovered and described a radically new method for the synthesis of ether in 1540, calling it "the sweet oil of vitriol."

At around the same time, Swiss physician Paracelsus discovered the anaesthetizing properties of ether.

STEAM TURBINE (1551)

The brilliant Turkish polymath Taqi al-Din described a functioning steam turbine, designed to power a rotating spit, in his wonderfully titled 1551 opus, *The Sublime Methods in Spiritual Devices*.

PENCIL (1560)

In the mid 1560s, the residents of a small village in England's Cumbria region stumbled across a massive deposit of graphite. The community first began using the substance to mark their cattle and sheep, and ultimately hit upon wrapping a wood casing around the graphite. It would take another two hundred years for the device to be completed with the invention of the eraser.

MERCATOR MAP PROJECTION (1569)

Flemish mapmaker Gerard Mercator developed the Mercator projection, a cartographical depiction of the world that allowed navigators to follow rhumb lines between two locations, thus accounting for compass bearing.

SUPERNOVAS AND COMETS (1572–1577)

The Danish nobleman Tycho Brahe's observation of a new star forming in 1572, and his detailed proof that the supernova was not changing position relative to other stars, undermined the prevailing orthodoxy that held that the heavens were incapable of change. Several years later, Brahe's equally precise observations of a comet showed that the object was farther away from the moon, and thus not part of earth's atmosphere.

STOCKING FRAME (1589)

English clergyman William Lee created the first working version of a stocking frame, a mechanical knitting machine used in the textile industry to mimic the motions of hand-knitting. Following the inventor's death, one of his assistants made a number of improvements on the device that much improved its functionality.

COMPOUND MICROSCOPE (1590)

Though a definite consensus does not exist on who invented the compound microscope, most historians credit either the Dutch spectacle maker Zacharias Janssen and his son Hans, or the German optician Hans Lippershey. In 1609 Galileo re-formed Janssen's original design into a more efficient machine. In the 1670s, Antoni van Leeuwenhoek first applied the microscope to the field of biology.

FLUSH TOILET (1596)

A water flushing device was invented in the late sixteenth century by Sir John Harrington, who installed a functioning version for his godmother, Queen Elizabeth, at Richmond Palace. But the device didn't take off until the late 1700s, when a watchmaker named Alexander Cumings and a cabinetmaker named Joseph Bramah filed for two separate patents on an improved version of Harrington's design.

PLANETARY MAGNETISM (1600)

English scientist William Gilbert realized that the earth itself was a magnet, a discovery first published in his treatise "On the Magnet" in 1600. Gilbert concluded that it was the earth's magnetic nature that allowed the compass to aid navigators. The nature of magnets had been studied by Aristotle and the ancient Chinese among others throughout history.

TELESCOPE (1600–1610)

A classic example of collective invention, the first telescopes and spyglasses began to appear in Europe in the first decade of the seventeenth century. Two patent applications were filed on designs in the Netherlands in 1608, and by 1609 Galileo was using a device he built with 20x magnification to gaze at the stars, discovering Jupiter's moons in the process.

ELLIPTICAL ORBITS (1605–1609)

The German astronomer and mathematician Johannes Kepler was the first to document the elliptical orbit that the planets took around the sun, though he built his equations by analyzing data collected by Tycho Brahe, his friend and occasional employer.

JUPITER'S MOONS (1610)

With the aid of a telescope, Galileo Galilei first observed the orbiting moons of Jupiter, thus proving the fundamental principle of the Copernican system, that the universe did not revolve around earth. Another scientist, Simon Marius, claimed to have discovered the moons five weeks prior to Galileo, but he never published his observations.

FLINTLOCK (1610)

French courtier Marin le Bourgeoys introduced the first fully developed flintlock mechanism to King Louis XIII in 1610, a device that became standard in firearms until the early nineteenth century. But Marin le Bourgeoy's discovery integrated many early innovations in firing mechanisms, from the matchlock to the snaphance.

SUNSPOTS (1610)

Sunspots, darkened, magnetic spots on the surface of the sun, were first observed almost simultaneously by a number of astronomers with the use of telescopes. Credit is alternately attributed to Galileo Galilei, Thomas Harriot, and Johannes and David Fabricius.

LOGARITHMS (1614)

In an effort to simplify the process of multiplying large numbers, mathematician John Napier conceived of logarithms as a way to express a number as a base raised to a power, e.g., 100 as 10^2, or 10×10. Logarithms have gone on to play an essential role in science and engineering.

BLOOD CIRCULATION (1628)

English physician William Harvey correctly theorized the movement of blood through the human body as pumped by the heart and cycled perpetually, dispelling earlier arguments for the existence of two separate circulation systems.

VERNIER SCALE (1631)

The Vernier scale, invented by French mathematician Pierre Vernier, can be used in conjunction with a larger scale to precisely measure extremely small units of space. It became widely employed in navigation systems.

OCEAN TIDES (1632)

Following in the steps of the ancients, Galileo Galilei ventured an explanation of ocean tides in relation to the sun. Johannes Kepler correctly theorized that it was the earth's relation to the moon that created the phenomenon, and Isaac Newton furnished the scientific community with a fully developed explanation in 1687.

SLIDE RULE (1632)

William Oughtred is commonly credited with inventing the earliest version of the slide rule, two parallel logarithmic scales that one could slide in relation to each other to conduct advanced calculations easily and quickly. Oughtred improved upon the design of a more basic model developed by Edmund Gunter as well as earliest conceptions by Galileo Galilei and John Napier.

LAW OF FALLING BODIES (1634)

For at least two thousand years, the Aristotelian consensus held that heavier bodies fall faster than lighter ones, until Galileo devised several ingenious experiments and formulated a mathematical equation to describe what we now call uniform acceleration. While several observational accounts predate Galileo's work, his account was the first definitive proof.

ANALYTIC GEOMETRY (1637)

French philosopher and mathematician René Descartes invented the system now known as analytic geometry as a way to express geometric shapes and properties with a coordinate system. By translating geometric structures, both two and three dimensional, into numerical representations, mathematicians could study and investigate them algebraically. Analytic geometry would later form one of the foundations of Isaac Newton's development of calculus.

BAROMETER (1643)

The barometer, a device designed to measure air pressure, grew out of Italian physicist Evangelista Torricelli's efforts to aid his mentor, Galileo, in an attempt to help miners pump water out of wells. While working with mercury, a heavier liquid than water, Torricelli discovered that variations in the height of mercury trapped in a tube from day to day were due to changes in the air's atmosphere. However, historians speculate that mathematician Gasparo Berti may have unwittingly invented a barometer a few years earlier.

MECHANICAL CALCULATOR (1645)

French mathematician and philosopher Blaise Pascal invented what is now called Pascal's Calculator, one of the most important precursors to the modern calculator, a device that could add and subtract through the use of spinning metal wheels stamped with numbers 0 through 9. While Pascal was the first to present his fully functioning invention to the public, a similar device had been conceived and developed by German Wilhelm Schickard, based on work by John Napier.

VACUUM PUMP (1654)

Like the barometer, the vacuum pump grew out of scientists' efforts to improve upon the capabilities of a suction pump. Through a series of experiments, Otto von Guericke discovered that it was possible to extract air or water from a sealed container, creating a vacuum. He demonstrated this principle before Emperor Ferdinand III by showing that two horses could not pull apart two bowls between which a vacuum had been created. Von Guericke drew on the work of Evangelista Torricelli, and his own work was improved upon by Robert Boyle and Robert Hooke.

PENDULUM CLOCK (1656)

Once again building on the ideas of Galileo, Dutch scientist Christiaan Huygens invented the most accurate clock to date by utilizing the regular oscillations of a weighted pendulum, regulated slightly by a mechanical device.

BALANCE SPRING WATCHES (1660)

Vastly improving upon the accuracy of earlier timepieces, a balance spring mechanism controlled the speed of the separate pieces of a watch with the help of a regulator, which ensured that the whole mechanism remained as consistent as possible. Robert Hooke and Christiaan Huygens are both credited with the invention—Thomas Tompion engineered the most effective regulator of the time around 1680.

BOYLE'S LAW (1662)

Boyle's law, developed by scientist Robert Boyle, states that given a fixed temperature and a closed system, the pressure and volume of gas will remain inversely proportionate; i.e., as one decreases, the other increases, in proportionate degrees. Boyle's assistant Robert Hooke assisted in the discovery of this law. French chemist Edme Mariotte discovered the same principle at roughly the same time, but Boyle published it first.

LIGHT SPECTRUM (1665)

Correcting earlier views that prisms colored light, Sir Isaac Newton demonstrated through a series of experiments that a ray of sunlight, shined through a prism, contained colors and was not colored by the prism, which only split the ray into its constituent parts. By isolating one color expressed by a prism, and shining it through yet another prism, Newton showed that the color remained consistent, and that the prism did not affect the shade.

MICROORGANISMS (1674–1680)

Thanks in part to his own improvements to the technology of the microscope, the Dutch scientist Antoni Philips van Leeuwenhoek was the first person to directly observe single-celled organisms, called "animalcules" at the time.

SPEED OF LIGHT (FIRST QUANTITATIVE MEASURE) (1676)

While Galileo had been able to establish that light traveled faster than sound, Danish astronomer Olaus Roemer, trying to account for disparities in his observations of eclipses, realized that the culprit was the amount of time light took to travel through space. By dint of advanced astronomical calculations, Roemer was able to approximate a speed of light not far off from modern estimates.

HOOKE'S LAW (1676)

Otherwise known as the law of elasticity, English scientist Robert Hooke discovered that the displacement or deformation of an object was proportionate to the amount of force exerted upon it—in other words, a spring stretches in proportionate amount to the degree of stress placed on it, before resuming its original shape.

PRESSURE COOKER (1679)

French physicist Denis Papin invented what he termed a steam digester—a sealed device containing liquid, which, when heated, created pressure within the closed unit, therefore raising the boiling point of the liquid, allowing for faster cooking times.

CALCULUS (1684, 1693)

Though the principles of modern calculus had been noted through the centuries, most historians credit Isaac Newton and Gottfried Wilhelm Leibniz with systematizing the methods and principles on a larger scale than had ever before been accomplished. Broadly described as a branch of mathematics that explains the principles of physics, Newton and Leibniz both lay claim to its invention, though history has since shown that both mathematicians arrived at many of the same conclusions independently, though with different systems of notation.

LAW OF UNIVERSAL GRAVITATION (1686)

While the story of Newton's apple may be the canonical example of private inspiration, the actual origins of the law are much murkier, including a famous battle between Robert Hooke and Newton over who first noted the inverse square relationship that governed the gravitational attraction between two objects.

THREE LAWS OF MOTION AND ORBITS OF COMETS (1687, 1705)

Newton's three laws of motion were first published in his groundbreaking *Philosophiæ Naturalis Principia Mathematica* in July 1687. Newton's friend and publisher, Edmund Halley, would then rely on those laws in producing the first accurate prediction of a comet's orbit around the earth.

PIANO (1700S)

Employed by the Medici court, Bartolomeo Cristofori sought to improve upon the harpsichord and clavichord by creating a similar instrument that would allow the player both expressive control and a larger spectrum of volume. He called it a "pianoforte," which has since been shortened to "piano."

TUNING FORK (1711)

Designed by the British musician John Shore, the tuning fork, or "pitch-fork," produced a very pure tone by which instruments could be accurately tuned.

STEAM ENGINE (1712)

Expanding upon the earlier, more primitive inventions of Denis Papin and Thomas Savery, Thomas Newcomen, an English blacksmith, utilized atmospheric pressure to propel a piston upward and downward by condensing steam, allowing an engine to pump water out of wells. It was the first commercially successful device of its kind.

MERCURY THERMOMETER (1714)

While crude thermometers were conceived by both Galileo Galilei and Isaac Newton, German physicist Daniel Gabriel Fahrenheit invented the first fully functioning mercury thermometer: a glass tube containing mercury that registered temperature according to the degree of heat applied to it, demarcating both the boiling and freezing temperatures of water.

OCTANT (1730)

Invented at about the same time but independently by Thomas Godfrey and John Hadley, the octant was a navigational device that spanned 45 degrees and, with the help of attached mirrors and a small telescope, could allow sailors to orient themselves at sea.

FLYING SHUTTLE (1733)

An invention that helped spur the Industrial Revolution, the flying shuttle was a device that sped up the process of weaving with a loom and required less manpower. The device did not become widely used until after inventor John Kay's death.

LINNEAN TAXONOMY (1735)

While the modern taxonomic scheme for organizing life still bears the name of the Swedish botanist and zoologist Carl Linnaeus, his model built on classificatory systems that had been evolving for hundreds of years. But Linnaeus did make several essential additions, most importantly the practice of naming each organism using a binomial structure, as in *homo sapiens*.

CHRONOMETER (1735)

Though countless versions of a chronometer had been developed since the early sixteenth century, the most fully realized device was created by the carpenter Thomas Harrison. The chronometer allowed navigators at sea to determine longitude and latitude by providing an accurate representation of time at a particular location.

LIGHTNING ROD (1750)

Ben Franklin first proposed the idea of a lightning rod in a letter written in 1750, and his descriptions were ultimately translated into French. The first test of Franklin's theoretical design was actually implemented in France in 1752.

SPINNING JENNY (1764)

A longstanding debate questions whether James Hargreaves was the true inventor of the spinning jenny, a machine that greatly improved the efficiency of the cotton industry. Some evidence suggests that Hargreaves was merely improving the design of an artisan named Thomas Highs. What is clear is that the Hargreaves design was greatly improved upon in the years following the production of his first model by weavers throughout Northern England.

CARBONATED WATER (1767)

Clergyman Joseph Priestley discovered that by charging water artificially with carbon dioxide, he could create an effervescent beverage, known today as seltzer. Though Priestley never capitalized on the business opportunities, many inventors after him did.

PHOTOSYNTHESIS (1770–1800)

Most commonly associated with the Austrian physician Jan Ingenhousz, the mechanism of photosynthesis was uncovered piece by piece over a thirty-year period, starting in 1770 with a series of experiments and essays written by Joseph Priestley. The cycle of carbon dioxide consumption and oxygen release, triggered by sunlight, wasn't fully formulated until the mid-1790s by the Swiss naturalist Jean Senebier.

PLANT RESPIRATION (1772–1773)

While Joseph Priestley is conventionally associated with the isolation of oxygen, he deserves more recognition for his discovery of plant respiration, circa 1773, which he collaborated on via post with his good friend Benjamin Franklin.

OXYGEN (1772–1776)

One of the great stories of scientific collaboration and rivalry, oxygen was isolated by three scientists in the mid-1770s: the Swedish chemist Carl Wilhelm Scheele; the British polymath Joseph Priestley, who named it "dephlogisticated air," after the reigning, and inaccurate, theory of phlogiston; and Antoine Lavoisier, who gave the element its name.

BIFOCALS (CIRCA 1780)

While the exact date of his invention in unclear, by the mid-1780s, Benjamin Franklin was writing to friends about how happy he was with his "invention of double spectacles, which serving for distant objects as well as near ones, make my eyes as useful to me as they ever were."

STEAMBOAT (1780–1810)

Robert Fulton is conventionally heralded as the inventor of the steamboat, but in fact Fulton was merely the first to turn the steamboat into a commercial success. A number of working steamboats had been built by engineers like John Fitch and James Rumsey over the preceding two decades.

MANNED HOT AIR BALLOON (1783)

While hot air balloons date back to first-century A.D. Chinese culture, the first manned flight was designed by the French entrepreneurs Joseph-Michel and Jacques-Etienne Montgolfier.

MILKY WAY (1785)

Many astronomers and scientists, including Abū Rayhān al-Bīrūnī and Galileo, contributed to the notion of the Milky Way as a collection of stars, but the first attempt to map the shape of the Milky Way was executed by William Herschel and his sister Caroline in 1785.

POWER LOOM AND COTTON GIN (1785, 1793)

The English clergyman Edmund Cartwright patented a power loom design in 1785, but, like Eli Whitney's cotton gin, the device relied on many subsequent improvements by other engineers for it to revolutionize the textile industry.

SMALLPOX VACCINE (1796)

The process of inoculating humans with small doses of the smallpox virus, usually using scabs from the skin of a victim, was practiced widely through China, Persia, and Africa after 1500. But the British scientist Edward Jenner was the first to design a vaccine based on a related cowpox virus that produced immunity to smallpox with much lower mortality.

LITHOGRAPHY (1796)

A playwright, Alois Senefelder struggled to find a way to distribute his writings cheaply—he eventually discovered that he could etch on a copper plate with acid and a needle. He improved this method, using the same fundamental idea, and called it "stone printing," which soon spread throughout Europe and the United States.

ELECTRIC BATTERY (1800)

The Italian count Alessandro Volta created the first battery out of zinc and copper discs, inspired by an argument with his peer Luigi Galvani, who believed that electricity emerged out of animal tissue.

ATOMIC THEORY (1800–1810)

While it drew heavily from the revolution in chemistry spearheaded by Antoine Lavoisier, the first rigorous argument that elements were made up of unique atoms of a distinct character was put forth by the English chemist John Dalton the first decade of the nineteenth century.

MOLECULES (1800–1810)

The notion that atoms form larger compound units, the most elemental of which is a molecule, was first formulated in the decades around 1800, and drew upon the related theories of the French chemist Joseph Proust, John Dalton, and the Italian count Avogadro.

SUSPENSION BRIDGE (1800–1830)

While numerous crucial improvements were added over the first half of the nineteenth century, the first functioning suspension bridge large enough to transport humans and horses was the Jacob's Creek Bridge, built in the early 1800s by the American judge and amateur engineer James Finley.

STEAM LOCOMOTIVE (1805)

A number of engineers built working steam-powered vehicles, some designed to run on roads, some on rails, in the last decades of the eighteenth century, though historians generally consider the train designed by Richard Trevithick in Wales in 1805 to be the first fully functioning steam locomotive.

PUNCH CARDS (1805)

The idea of using punch cards for programming mechanical looms is generally credited to Joseph Marie Jacquard, but weavers in the early 1700s, including Basile Bouchon and Jean Falcon, experimented extensively with punch card control of warp threads.

SPECTROSCOPE (1814)

German lenscrafter Joseph von Fraunhofer invented the spectroscope, a device that measures properties of light, in order to study dark lines occurring in various forms of spectra, which he later discovered were areas of the spectrum where light is absorbed.

STETHOSCOPE (1816)

A French physician named René Laennec invented the stethoscope after improvising one with a roll of paper while treating a woman suffering from heart disease.

BICYCLE (1817–1863)

The first two-wheeled, steerable vehicle was designed by a German baron named Karl von Drais, and mimicked by dozens of entrepreneurs throughout Europe in the following decade. But it wasn't until the 1860s that pedals and rotary cranks were added to the device.

BRAILLE (1821)

Louis Braille, a blind, French fifteen-year-old, invented Braille—a tactile form of reading—by improving upon a more rudimentary system of raised bump, tactile text (night-writing) conceived by an army captain, Charles Barbier.

ELECTRIC MOTOR (1821–1850)

More than a dozen scientists and entrepreneurs contributed to the design of the electric motor in the first half of the eighteenth century, beginning with the English chemist and physicist Michael Faraday's demonstration, in 1821, of a system for converting electrical energy into mechanical energy.

SECOND LAW OF THERMODYNAMICS (1824)

The second law of thermodynamics, which evolved over the years in the hands of various scientists, including Sadi Carnot and Rudolf Clausius, states a theory of universal entropy that invalidates the possibility of perpetual motion machines.

GEOLOGICAL UNIFORMITARIANISM (1830)

The idea that the geological state of the earth was based on consistent forces acting over very long time scales is largely attributed to Charles Lyell's *Principles of Geology*, published in 1830, though the term itself comes from a review of Lyell's book written by William Whewell. Lyell's ideas would subsequently form the platform on which Darwin based his biological theory of evolution.

CHLOROFORM (1831)

Chloroform, a colorless, organic compound, was discovered at about the same time by three different scientists in three different countries—Eugene

Soubeiran, Samuel Guthrie, and Justus von Liebig. It was used as a treatment for asthma and as a powerful alternative to ether as an anesthetic.

REFRIGERATOR (1834)

After acquiring a patent for a vapor-compression refrigeration system, mechanical engineer Jacob Perkins invented the first practical refrigerator in 1834, though an earlier refrigeration machine had been invented by American inventor Oliver Evans in 1805.

REVOLVER (1836)

Improving upon the flintlock firing mechanism, in 1836 American inventor Samuel Colt designed and patented the revolver, a handgun that featured a rotating cylinder with multiple chambers for bullets.

PROGRAMMABLE COMPUTER (1837)

Although a working version was never built, Charles Babbage outlined the basic principles of the programmable computer—including the notions of what we now call software, CPU, and memory—in his legendary Analytical Engine, which he first published a description of in 1837. Lord Byron's daughter Ada Lovelace wrote the first computer algorithm for the device.

TELEGRAPH (1838)

In an effort to improve clumsier, five-wire models of the telegraph, inventor Samuel Morse and his assistant Alfred Vail created a one-wire model that used electric signals to shift an electromagnet in a patterned print across paper, known as Morse code.

PHOTOGRAPHY (1839)

Most historians credit French chemist Louis Daguerre with developing the first practical photographic process, which involved fixing images on copper places covered in a chemical substance by exposing them to light. Daguerre's methods were deeply influenced by the innovations of Frenchman Joseph Nicephore Niépce.

VULCANIZED RUBBER (1839)

After years of trial and error, Charles Goodyear discovered vulcanized rubber—which unlike natural rubber, maintained its shape despite exposure to pressure and heat—almost by accident, and fought for the rest of his life to claim royalties on the product. Not long after Goodyear's discovery, Thomas Hancock beat him to the patent.

SEWING MACHINE (1845)

The invention of the modern, practical sewing machine was largely due to the individual innovations of two men, American mechanic Elias Howe, who developed the machine's lockstitch mechanism, and American inventor Isaac Singer, who pioneered the vertical motion mechanism for the needle. The two men would clash over credit for the invention.

NITROGLYCERINE (1846)

Working as an assistant to professor J. T. Pelouze, Ascanio Sobrero first discovered and synthesized nitroglycerine—aware of its explosive potential, Sobrero warned against incautious use of the chemical and at times even seemed to regret its discovery.

ABSOLUTE ZERO (1848)

Drawing on the work of earlier scientists studying temperature, Kelvin developed absolute zero, which forms the lowest point of his Kelvin scale, representing the point at which all matter ceases to move—roughly 273.15° C.

PIG IRON/STEELMAKING (1850–1860)

Though the process of making steel would continue to be improved for years after Henry Bessemer's innovations, the American inventor discovered the first means to mass-producing steel. By oxidizing pig iron, Bessemer was able to manufacture comparably high-quality steel in large quantities, eventually aiding the construction of skyscrapers.

ELEVATOR (1853)

While rudimentary versions of "lifts" had existed since the Middle Ages, American inventor Elisha Otis sparked wide public use of such machines in 1853 by developing a safety brake, following the introduction of steam and hydraulic elevators around 1850.

ASPIRIN (1897)

While the pain-relieving properties of willow bark, whose medicinal quality derived from the tree's salicin, had been understood and prescribed since Hippocrates, the drug's use had been plagued by side effects, primarily stomach pains. French chemist Charles Gerhardt discovered that adding sodium and acetyl chloride assuaged the intestinal irritation, making for a better drug.

BUNSEN BURNER (1856)

German chemist Robert Bunsen developed the burner in order to carry out experiments on spectral emissions of elements, for which the technology did not yet exist. Stymied by the weak gas burners of the day, Bunsen produced a burner with an incredibly hot and nearly invisible flame, and it became the standard laboratory burner many still use today.

MASON JAR (1858)

Improving upon the inefficient jars commonly used at the time, tinsmith John L. Mason invented a type of jar that would one day bear his name: a blocky glass container with a screw top and rubber lining to create an airtight seal. The Mason jar became essential in preserving perishable goods.

LEAD-ACID BATTERY (1859)

French physicist Gaston Plante invented the first rechargeable battery while experimenting with the conductive power of rolled sheets of lead and sulfuric acid.

NATURAL SELECTION (1859)

Natural selection was first formulated by Charles Darwin in the late 1830s, though he did not publish his ideas until 1859 in his book *The Origin of Species*, after being spurred on by the very similar theories that had been independently developed by the British naturalist Alfred Russel Wallace.

GATLING GUN (1861)

Laboring under the belief that a revolving machine gun would create less bloodshed on battlefields by reducing the number of soldiers needed, inventor Richard Gatling created the Gatling gun, a hand-cranked continuously and rapidly firing weapon drawn on two wheels.

VACUUM CLEANER (1861)

Though many inventors created versions of what we know today as a vacuum cleaner in the late nineteenth century and early twentieth century, Ives W. McGaffey patented the first manually powered vacuum cleaner—or "sweeping machine"—in 1861, marketing the device's ability to clean carpets.

PLASTIC (1862)

British metallurgist Alexander Parkes developed the first major commercial man-made plastic—a synthetic material made from cellulose and treated with nitric acid—and debuted it at the 1862 World's Fair in London. Improvements were made on the material through the nineteenth and twentieth centuries.

GERM THEORY (1862)

While the idea that germs carried contagious disease was not new and had been proposed before, French chemist Louis Pasteur was one of the first to develop experiments to prove the theory conclusively.

DYNAMITE (1863)

Seeking to develop new methods for blasting rock more effectively, Swedish industrialist Alfred Nobel built on his experiments with nitroglycerin and invented a detonator that used a strong shock to spark explosions, which he patented in 1863.

PERIODIC TABLE (1864)

In 1864, Russian chemist Dmitri Mendeleev developed upon earlier notions of British chemist John Newlands that chemical elements could be arranged in a pattern according to their atomic masses, providing a more comprehensive chart with a focus on recurring trends in properties.

DISCOVERY OF BENZENE STRUCTURE (1865)

Following the discovery of benzene in 1825, German chemists Joseph Loschmidt and August Kekule von Stradonitz theorized a similar structure of the organic chemical compound—a ring of six carbon atoms with alternating single and double bonds. Kekule's discovery was inspired by his legendary dream of the tail-swallowing serpent.

HEREDITY (1865)

The idea that parents pass certain hereditary qualities to their offspring was originated by Augustinian monk and scientist Gregor Mendel from his work on plants, though his principles were synthesized into a wider theory of genetics by Thomas Hunt Morgan in the early twentieth century.

TYPEWRITER (1868)

After the invention of an inefficient typographer machine in 1829, American inventor Christopher Latham Sholes patented the first practical typewriter in 1868 with the help of his associates, pioneering a type-bar system and the QWERTY arrangement of keys to avoid jamming.

TELEPHONE (1876)

The patent for the invention of the telephone was a hotly contested item, leading to a last-minute race to the patent office between American engineer Alexander Graham Bell and American electrical engineer Elisha Gray. Bell ultimately received the patent for the device, which transmitted voice signals electrically.

ENZYMES (1878)

First named by German doctor Wilhelm Kühne in 1878, enzymes—proteins that act as catalysts for chemical reactions by speeding up the process—were

more fully understood due to the studies of German chemist Eduard Buchner and French chemist Louis Pasteur.

LIGHTBULB (1879)

By using electricity to heat a filament, causing it to glow and create light,
American inventor Thomas Alva Edison is often considered the inventor of
the lightbulb, replacing gas lighting as the main source of illumination. But
Edison's work built on the designs of at least a half dozen other inventors
who went before him, including Joseph Swan and William Sawyer.

CELL DIVISION (1879)

The discovery of cell division, the process known as mitosis among eukary-
otes in which a parent cell divides into daughter cells, was the joint discovery
of German biologist Walther Flemming, Eduard Strasburger, and Edouard
van Beneden.

SEISMOGRAPH (1880)

Hired by the Japanese government to study tremors and earthquakes, three
British scientists worked on creating a device that could measure and classify
the strength of earthquakes, now known as a horizontal seismograph, characterized by its use of a pendulum. Of the three, John Milne generally re-
ceives the lion's share of credit for the invention.

INFANT INCUBATORS (1881)

Inspired by the use of an incubator for baby birds, French obstetrician
Étienne Stéphane Tarnier began putting infant incubators—heated cribs for
newborns—into regular practice in hospitals. The original designs for the
infant incubator were created by French surgeon Jean-Louis-Paul Denucé
and German gynecologist Carl Credé.

WELDING MACHINE (1885)

Russian inventors Nikolai Bernardos and Stanislav Olszewski created the first electric arc welder in 1885, though the principle underlying welding—that the application of heat can be used to join metallic pieces—had been understood and utilized for centuries.

MOTORCYCLE (1885)

German inventor Gottlieb Daimler expanded upon Nikolaus Otto's internal-combustion engine by connecting it to a bicycle, thus powering the vehicle by gas, not manpower. A steam-engine device resembling a modern motorcycle was invented in 1867.

AUTOMOBILE (1885)

In roughly the same year, 1885, German engineer Gottlieb Daimler and Wilhelm Maybach created a four-wheel, four-stroke engine automobile and German engineer Karl Benz, who most historians credit as the ultimate inventor of the modern automobile, designed a motor car powered by an internal combustion engine and gasoline.

INDUCTION MOTOR (1885)

Both Italian physicist Galileo Ferraris and Austrian inventor Nikola Tesla filed patents in the same year for the induction motor, an alternating current motor that functions via electromagnetic power.

CALCULATOR (1885)

Following centuries of attempts to develop a reliable calculating machine, American inventor William Seward Burroughs created a "calculating machine" in 1885 that formed the basis for all further improvements in calculators.

CONTACT LENSES [1887]

Though Leonardo da Vinci is said to have sketched the first designs for corrective vision lenses, German glassblower F. A. Muller first conceived of lenses that would replicate the shape of the human eye and improve vision. With the help of his assistants, German physicist Adolf Eugen Fick further improved the design by creating lenses that conformed to the eye more comfortably than any previous version.

EKG [1887]

The EKG evolved over a string of developments, though the most important contribution may have been Augustus Waller's in 1887; Waller was the first scientist to publish an EKG by attaching an electrometer to a projector.

MOTION PICTURE CAMERA [1888]

American inventor Thomas Edison patented one of the early versions of a motion picture camera—which he called a "kinetoscope"—but his device drew heavily on similar work done by English photographer Eadweard Muybridge and the discoveries of other experimenters with the photographic medium in the late nineteenth century.

MITOCHONDRIA [1890]

German pathologist Richard Altmann is generally credited with first discovering mitochondria—organelles that provide cells with the majority of their chemical energy—for postulating that they were fundamental units of cell activity. Numerous scientists continued to make large strides in their understanding of mitochondria throughout the twentieth century.

TESLA COIL (1891)

The Czech inventor Nikola Tesla invented the Tesla coil, a high-frequency transformer that creates extremely large amounts of voltage, and which was used commercially for lighting and in radio transmission.

RADIO (1896)

While Italian engineer Guglielmo Marconi is traditionally credited with the invention of modern radio by using radio waves to create a system of wireless telegraphy (he received a patent for the creation in 1896), the contributions of Nikola Tesla, Karl Ferdinand Braun, and Heinrich Hertz, among others, were essential to the final design.

RADIOACTIVITY (1896)

Expounding on the closely preceding discoveries of German physicist Wilhelm Roentgen and French physicist Henri Becquerel, Polish chemist Marie Curie and her husband, Pierre Curie, formed a theory of radioactivity, which describes the spontaneous disintegration of atomic nuclei.

ELECTRON (1897)

British physicist J. J. Thomson, aided by Irish physicist John Townsend and British physicist H. A. Wilson, discovered the electron while experimenting with cathode rays, proposing that they were composed of negatively charged particles smaller than atoms, which he called corpuscles, later renamed electrons.

BLOOD TYPES (1901)

In 1901, Austrian physician Karl Landsteiner published results of studies in which he argued that four distinct blood types existed—distinguished by the presence of particular antibodies and antigens—and that blood transfusion between two individuals could be successful only if they shared the same blood type.

AIR CONDITIONING (1902)

In an attempt to solve a printing factory's difficulties with the effects of fluctuating temperatures and humidity on paper, Willis Haviland Carrier conceived of a way to reverse the process of heating to create cold air, thus controlling the amount of moisture in the air. Carrier would go on to start a company dedicated to air conditioning, and it was not long before models were made available for domestic use.

STRATOSPHERE (1902)

German meteorologist Richard Assmann and French meteorologist Léon Teisserenc de Bort are both credited with the discovery of the stratosphere, the second layer of the earth's atmosphere, in 1902.

ENGINE-POWERED AIRPLANE (1903)

Inspired by the efforts of German aerial engineer Otto Lilienthal, the Wright brothers experimented with the flying patterns of kites and eventually developed the first engine-powered airplane, which succeeded in performing sustained flight in 1903.

VITAMINS (1905)

While people knew for centuries that eating certain foods could prevent disease, the English doctor William Fletcher discovered in 1905 that unpolished white rice was instrumental in creating immunity to beriberi disease, while other kinds of rice were not, leading him to believe there were nutrients in the unpolished rice whose absence in a person's diet would increase their susceptibility to disease.

HORMONES (1905)

Confirming earlier scientific work on internal secretions in the human body, English physiologists Ernest Henry Starling and William Maddock Bayliss showed that a chemical agent released in one part of the body could affect the functioning of another part of the body, via the bloodstream. The discovery of hormones would later lead to the invention of both oral contraceptives and insulin.

MASS-ENERGY EQUIVALENCE (1905)

Theoretical physicist Albert Einstein stated in a paper published in 1905 that the mass of a body is equivalent to its energy content, expressed in the famous equation $E=mc^2$, or energy equals mass times the speed of light squared.

SPECIAL RELATIVITY (1905)

The theory of special relativity—developed by Einstein in 1905—concerns the motion and behavior of particles moving at close to the speed of light, and is based on two postulates: that the speed of light is the same, regardless of the speed of the observer, and that the laws of physics are consistent when observed from any inertial, or nonaccelerating, frame of reference.

EARTH'S CORE (1906)

Irish seismologist Richard Oldham deduced that the earth's core was made of less dense, more liquid material than the rock surrounding it by studying why earthquake waves moved slower through the earth's core than through the mantle.

NEUROTRANSMITTERS (1906)

Spanish physician Santiago Ramón y Cajal revolutionized the theories of the structure of the nervous system in the early twentieth century, aided by methods developed by Italian physicist Camillo Golgi. Cajal's theory that the nervous system was composed of billions of tiny nerve centers—to become known as neurons—led the discovery of neurotransmitters, chemicals that relay messages across synapses.

WASHING MACHINE (1908)

American engineer Alva John Fisher pioneered the first electric washing machine by attaching a motor to the traditional model of a hand-cranked washer. The Chicago-based Hurley Machine Company introduced the product in 1908.

GENES ON CHROMOSOMES (1910)

American embryologist Thomas Hunt Morgan's experiments with genetic mutations and the fruit fly *Drosophila melanogaster* led him and his team of students at Columbia University to discover how heredity was in part governed by genes transported by chromosomes.

SUPERCONDUCTIVITY (1911)

In 1911, Dutch physicist Heike Kamerlingh Onnes tested the behavior and properties of metals such as lead, tin, and mercury when placed at liquid helium temperatures, and discovered that they lost all resistance when cooled to cryogenic levels. This quality became known as superconductivity.

COSMIC RAYS (1913)

The discovery of cosmic rays—particles that bombard earth from beyond its atmosphere—was the culmination of the work of a number of scientists in the early twentieth century, although the German physicist Werner Kolhörster did receive a Nobel Prize for his work and research in the nascent field. However, Kolhörster's experiments leaned heavily on earlier discoveries by Victor Hess and Theodor Wulf.

ELECTRON'S ROLE IN CHEMICAL BONDING (1913)

Danish physicist Niels Bohr proposed his model of the electron (loosely based on British chemist Ernest Rutherford's model) in 1913, postulating that electrons travel in patterned orbits around the nucleus of an atom, and further theorized that the chemical makeup of an element is derived from the number of electrons in the atom's orbit. Bohr's discovery revealed the electron's fundamental role in chemical bonding.

CONTINENTAL DRIFT (1915)

In 1915, German meteorologist and geologist Alfred Wegener published a book in which he argued that all the continents of the earth had once been part of one massive landmass called Pangea, which had slowly split apart over time. Wegener's ideas were initially rejected, but became universally accepted by the 1960s.

MOVING ASSEMBLY LINE (1913)

Heralding the era of mass production, the Ford Motor Company instituted a moving assembly line to construct cars under Ford's leadership in 1913, lowering the price of cars and quickening their production. The inspiration for the assembly line came from nineteenth-century midwestern meatpacking factories.

THEORY OF GENERAL RELATIVITY (1915)

Theoretical physicist Albert Einstein argued in 1915 that matter warps time and space, allowing large masses to bend light. One of the seminal aspects of this theory was Einstein's idea that the pull of gravity in one direction was equivalent to the force of acceleration in the opposite direction. Einstein's theory was proved in 1919 in a study of solar eclipses.

HELICOPTER (1920)

Many failed, but promising models of primitive helicopters preceded the type created by Argentinean inventor Raúl Pateras Pescara. Pescara's helicopter was the first to achieve cyclic pitch, or control of the rotor blades, and he set the world record in 1924 for flying close to a half mile in a little over four minutes.

QUANTUM MECHANICS (1925)

The field of quantum mechanics, the physics of atomic and subatomic scales, can be loosely dated back to 1925, when Werner Heisenberg published his first paper on the topic, but was largely created as a result of efforts of a number of innovative thinkers, including Einstein, Bohr, Planck, and others, working from the 1900s to the 1930s.

LIQUID ENGINE ROCKET (1926)

American physicist Robert H. Goddard overcame criticism of his belief in the future of rockets and helped pioneer the field in 1926, when he set off the first liquid-fueled rocket in a New England cabbage field.

UNCERTAINTY PRINCIPLE (1927)

First presented in a letter in 1927, German physicist Werner Heisenberg's uncertainty principle stated that the more precisely the position of a subatomic particle position was known, the less precisely one could know the particle's momentum. Interpreted in a number of ways, the most influential notion has been the idea that the act of observation changes the very object being observed.

TELEVISION (1927)

American inventor Philo Farnsworth filed a patent for the first complete electronic television in 1927, though technological developments (cathode ray tube, Audion vacuum tube) leading to this final stage were contributed by many engineers and inventors over the course of the previous century.

PENICILLIN (1928)

While healers dating back to ancient civilization realized that molds could be used to help cure infection, it was a famous mistake in Scottish biologist Alexander Fleming's laboratory that eventually brought penicillin to public attention as a miraculous antibiotic. The discovery occurred when a spore of *Penicillium notatum* floated into a petri dish containing mold, sparking Fleming's observation that the spore was inhibiting the growth of the bacteria.

EXPANSION OF THE UNIVERSE (1929)

While working at an observatory in California, American astronomer Edwin Hubble determined that the universe was expanding while measuring the redshifts (shifts in the frequency of photons) of distant galaxies and discovering that they were moving away from each other at a rate constant to the distance between them.

JET ENGINE (1930)

The credit for the invention of the jet engine is shared between German engineer Hans von Ohain and RAF officer Frank Whittle, who both independently developed the engine model, propelled by ignited, compressed air and based on the principles of Newton's third law of physics.

NEUTRON (1932)

British physicist James Chadwick discovered the neutron—a subatomic particle with no electrical charge—in 1932, putting in place some of the first steps toward the development of the atomic bomb.

RADAR (1935)

Scottish meteorologist Robert Watson-Watt drew on previous research using radio waves to sense inclement weather and successfully employed a short-wave radar in 1935 to detect a bomber in the air, a discovery that would prove instrumental in Britain's defense during the Battle of Britain.

TAPE RECORDER (1935)

Tape recorders began appearing in the early 1930s, led by German technology companies. German-born engineer Semi Joseph Begun developed the first consumer tape recorder, a "Sound Mirror," in 1935 by employing his research on magnetic recording, using a specially coated paper and plastic.

NYLON (1937)

While heading the research department at DuPont, American chemist Wallace Carothers developed nylon—"the miracle fiber," a man-made synthetic rubber—in part to create an alternative to silk, which at that time was difficult to obtain because of shaky trade relations with Japan.

ECOSYSTEM (1935)

First coined in 1935, "ecosystem" came to be fully defined in 1935 by British chemist Arthur Tansley as a natural system in which all physical and organic elements coexist and function as a more or less complete unit.

KREBS CYCLE (1937)

The Krebs cycle, the chemical mechanism by which a cell's respiratory system functions, was formulated by German biochemist Hans Krebs in 1937, building on extensive advances by multiple scientists over the preceding decade in understanding the way cells convert nutrients into energy.

ATOMIC REACTOR (1938)

Italian physicist Enrico Fermi and Hungarian physicist Leo Szilard created the first nuclear reactor in the 1930s, based on studies conducted by Fermi and his colleagues on beta decay and the theory of neutrons.

COMPUTER (1944)

German engineer Konrad Zuse is credited by many with inventing the first fully functioning modern computer, based on a binary system, in 1944. However, Charles Babbage, Alan Turing, and John Vincent Ansoff can all also be credited with inventing various forms of computers.

DNA AS GENETIC MATERIAL (1944)

The idea that DNA carries genetic material was initially established by the famous Avery-MacLeod-McCarthy experiment in 1944, which demonstrated that, since DNA could cause the transformation of bacteria, it could be seen to play a major role in hereditary transfer and the passing of genes from one generation to the next.

MICROWAVE OVEN (1946)

Percy Spencer, an American engineer, discovered the possibility of creating a microwave oven somewhat by accident when he noticed a candy bar melting while building a magnetron for Raytheon during an experiment with electromagnetic radiation.

TRANSISTOR (1947)

Enlisted to improve upon the vacuum tube, experimental theoretician Bill Shockley and physicists Walter Brattain and John Bardeen experimented with semiconductors at Bell Labs, eventually producing a reliable transistor that could amplify and switch electronic signals.

RADIOCARBON DATING (1949)

While at the University of Chicago, American physicist Willard Libby worked with his colleagues to develop radiocarbon dating, a method of determining the age of organic substances by how much carbon-14 is present in the material, revolutionizing the field of archeology.

ARTIFICIAL PACEMAKER (1950)

Although a few primitive versions of artificial heart pacemakers had been designed before 1950, Canadian engineer John Hopps is generally credited with the invention of the device, which uses electrical impulses to regulate and simulate the normal beating patterns of the heart. Internal pacemakers would not be developed until 1958.

ORAL CONTRACEPTIVE (1951)

A group of loosely connected scientists, most prominently Harvard professor John Rock, developed the birth control pill in the early 1950s, funded in part by the American birth control advocate Margaret Sanger. Leading a research group at the pharmaceutical company Syntex, American chemist Carl Djerassi worked on developing a steroid hormone, cortisone, which eventually led to the synthesis of norethindrone, a progestogen, which became a fundamental part of the first successful oral contraceptive.

EARLY LIFE SIMULATED (1953)

In an effort to understand the conditions governing early life on earth, American chemist Stanley Miller and American physical chemist Harold Urey created a closed system, including the elements they believed were present in earth's early atmosphere such as hydrogen, methane, and water. Miller and Urey discovered that amino acids could be easily produced under such conditions.

DOUBLE HELIX (1953)

Drawing on previous studies of nucleotides in DNA, American molecular biologist James D. Watson and British molecular biologist Francis Crick experimented with models of different combinations of nucleotides using paper and wire, and eventually settled upon the intertwined, dual, nucleotide strands that we now recognize as the double helix.

VCR (1956)

The invention of the VCR, or video cassette recorder, is generally attributed to the American engineer Charles Paulson Ginsburg, who developed the device while at the Ampex Corporation by applying high-frequency signals onto magnetic tape.

LASER (1958)

While at Bell Labs, American physicists Arthur L. Schawlow and Charles H. Townes began intensive investigation of infrared or visible radiations, initially developing what they called a maser, which would later evolve into "light amplification by stimulated emission of radiation," or a laser.

GPS (1958)

GPS, or Global Positioning System, a navigational system that uses satellites as reference points to calculate geographical positions, was developed by the American engineer Ivan Getting and his team at the Raytheon Corporation, at the behest of the U.S. Department of Defense, after the initial foundational work of Guier and Weiffenbach tracking the orbit of *Sputnik* in 1957.

COSMIC MICROWAVE BACKGROUND RADIATION (1965)

While working with receiver systems at Bell Labs, American astronomers Arno Penzias and Robert Woodrow Wilson were confounded with a sound they could not identify, which they ultimately realized was cosmic microwave background radiation, a remaining radio trace of the Big Bang.

PULSARS (1967)

Pulsars—pulsating neutron stars that appear to blinking—were observed and discovered in 1967 by Jocelyn Bell Burnell, a graduate student working under the British astronomer Antony Hewish, who would later receive a Nobel.

RNA ALSO GENETIC (1967)

Mirroring previous discoveries that DNA carried genetic material, American microbiologist Carl Woese theorized in 1967 that RNA, ribonucleic acid, could store information such as genes, and may have played a role in the development of early and precellular life.

RESTRICTION ENZYMES (1968)

First isolated in 1968 by geneticists H. O. Smith, K. W. Wilcox, and T. J. Kelly at Johns Hopkins University, restriction enzymes are found in bacteria and can cut DNA at specific sequences, thus paving the way for the future of recombinant DNA molecules.

GRAPHICAL USER INTERFACE (1968–1974)

The use of visual metaphors to represent data on a computer screen, along with the concept of a mouse as pointing device, dates back to a legendary demo by the Stanford professor Douglas Engelbart. Elements of the GUI were also evident in Ivan Sutherland's 1963 program Sketchpad. The idea was refined and expanded by the Xerox PARC lab in the early 1970s.

INTERNET (1970–1975)

Assisted by many other computer scientists, the American Vinton Cerf designed and created the original model of the Internet, building on his early research and experiments with packet-switching networks, supported by the U.S. Department of Defense Advanced Research Projects Agency.

CT SCAN (1971)

Using a grant provided by the British Department of Health and Social Services, British electrical engineer Godfrey Hounsfield conceived and designed the first CT scan (computerized axial tomography), which sent multiple X-ray beams through the human body, providing a near three-dimensional image.

MRI (1974)

Building on the discoveries of early MRI inventors, Raymond Damadian discovered that magnetic resonance imaging would command different responses from cancerous and noncancerous animal tissue.

ENDORPHINS (1975)

Discovered at about the same time by two research teams working independently, endorphins were first described when American scientist John Hughes and German-born British biologist Hans Kosterlitz published their results of a study in which they removed an amino-acid molecule from the brain of a pig, which they believed would bolster investigations of the brain's receptors for morphine.

PERSONAL COMPUTER (1976)

Legendarily working out of a garage, entrepreneurs and college dropouts Steve Wozniak and Steve Jobs designed one of the first personal computers, or microcomputer—Apple I—in 1976, creating the first single-circuit board computer, though many important models, including the Altair, preceded it.

ONCOGENES (1976)

Bolstering the understanding of cancer and how malignant tumors are created, American immunobiologist J. Michael Bishop and cellular biologist Harold Varmus discovered the first human oncogene in 1970.

RNA SPLICING (1977)

British biochemist Richard J. Roberts and American Phillip A. Sharp share both the credit and the Nobel Prize for their independent discoveries of gene-splicing—the removal of introns—though some controversy arose over lack of acknowledgment of Roberts's colleagues.

ARCHAEA (1977)

Realizing that a number of organisms did not fit into the traditional categorization of plant or animal, American microbiologist Carl Woese and his colleagues created a new classification of life, archaebacteria, shortened to archaea, to accompany bacteria and fungi.

GLOBAL WARMING (1970–1980)

While theories had been proposed throughout the twentieth century suggesting that carbon dioxide buildup could lead to a warmer planet, the science of global warming did not reach critical mass until the 1970s and 1980s, as a broad network of scientists, working in multiple fields, began to track and model changes in the earth's atmosphere.

ASTEROID EXTINCTION (1980)

On the basis of substantial geological evidence, scientific father-son team Luis and Walter Alvarez theorized in 1980 that 65 million years ago, a giant asteroid had struck earth, killing the dinosaur population.

DNA FORENSICS (1984)

British geneticist Alec Jeffreys discovered DNA forensic fingerprinting by accident while looking at an X-ray from a DNA experiment that appeared to show the variations in the DNA of his technician's family. Jeffreys soon after realized that DNA fingerprinting could be used to identify individuals by their genetic code. Dozens of other scientists refined Jeffreys's approach before it could be used in criminal cases.

UNIVERSE ACCELERATING (1988)

Based on observations of the stars created by white dwarf star explosions, High-Z Supernova team scientists led by astronomers Adam Riess and Brian Schmidt determined that the universe was expanding at an accelerating rate.

WORLD WIDE WEB (1989–1992)

British software engineer Tim Berners-Lee designed the program for the World Wide Web almost completely independently while working at CERN (European Laboratory for Particle Physics), in an attempt to create a "hypertext notebook," which was inspired by the memory of a childhood encyclopedia.

GAMMA RAY BURSTS (1997)

Gamma ray bursts—flashes of gamma rays coming from deep outer space—were first observed in 1967 by unclassified military satellites. The bursts befuddled scientists, uncertain of their nature or origin, until 1997, when the Italian-Dutch satellite BeppoSAX was able to target the burst position, leading scientists to understand that the rays were caused by residual X-ray emissions.

Notes and Further Reading

ON INNOVATION

An extensive literature exists on the question of innovation, particularly with reference to scientific and technological fields. I have tried to include a broad survey of these works in the bibliography, but several works have been disproportionately influential on my argument and method in this book. Dean Keith Simonton's *Origins of Genius* and Howard Gruber's *Darwin on Man* both explicitly take a Darwinian approach to innovation, and use that approach to make sense of Darwin's own distinct genius. Arthur Koestler's *Act of Creation* and Thomas Kuhn's *Structure of Scientific Revolutions* remain essential platforms for the understanding of new ideas. Richard Florida's *Rise of the Creative Class* looks at creativity in an urban context. Richard Ogle's *Smart World* explores the intellectual and physical context of idea formation, as does Howard Gardner's *Creating Minds*. Everett M. Rogers's *Diffusion of Innovations* is the canonical study of the way good ideas spread through organizations and society. Mihaly Csikszentmihalyi's *Flow* and *Creativity* explore the psychological states of intense creativity. The power of group and "end-user" innovation has been persuasively documented by Eric von Hippel in *Democratizing Innovation* and by Amar Bhidé's *Venturesome Economy*. And many of our clichés about the origins of good ideas are delightfully debunked in Scott Berkun's *Myths of Innovation*.

INTRODUCTION: REEF, CITY, WEB

The account of Darwin's voyage to the Keeling Islands is drawn from Darwin's own narrative in *Voyage of the* Beagle, as well as from some of the correspondence included in *The Autobiography of Charles Darwin*, and R. D. Keynes's *Charles Darwin's* Beagle *Diary*. The connection between Darwin's theories about coral reef formation and his later insights into the mechanism of natural selection is addressed in Howard Gruber's *Darwin on Man*. The original study on superlinear scaling in urban environments is available in "Growth, Innovation, Scaling, and the Pace of Life in Cities," by Bettencourt, et al. A thoughtful layperson's introduction to Kleiber's law and its application to urban culture can be found in George Johnson's "Of Mice and Elephants: A Matter of Scale." For a thorough history of HDTV's development, see Joel Brinkley's *Defining Vision*. An informative chart of twentieth-century technology adoption rates in the United States can be found at http://www.nytimes .com/imagepages/2008/02/10/opinion/10op.graphic.ready.html. John Cloud's "The Gurus of YouTube" offers a history of the company's founding. For a compelling overview of the Web's "generative" powers, see Jonathan Zittrain's *The Future of the Internet—And How to Stop It*. For more on the evolution of software interfaces, see Howard Rheingold's *Tools for Thought* and my *Interface Culture*. The notion of "patterns" of innovation is loosely based on the concept of patterns and metapatterns developed by Gregory Bateson in *Mind and Nature*. The "long zoom" approach is discussed in more detail in the appendices of my earlier books *Everything Bad Is Good for You* and *The Invention of Air*. The idea has roots in Edward O. Wilson's notion of "consilience," and was partially inspired by a "pace-layered" drawing of civilization that I first encountered in Stewart Brand's *How Buildings Learn*.

CHAPTER 1: THE ADJACENT POSSIBLE

For a history of the incubator, see Jeffrey Baker's "The Incubator and the Medical Discovery of the Premature Infant." The site Neonatology on the Web (http://www .neonatology.org/) maintains an excellent archive on the history of incubators and other neonatal technologies. For more on Design That Matters's approach to innovation, see Timothy Prestero's "Better by Design." Additional information on the NeoNurture device can be found at designthatmatters.org. Kauffman's theory of the adjacent possible is outlined in his book *Investigations*. The social causes of multiple simultaneous discovery are outlined in Ogburn and Thomas's "Are Inventions

Inevitable?" The phenomenon is also discussed at length in Dean Keith Simonton's *Creativity in Science*. For more on the discovery of oxygen, see Kuhn's *Structure of Scientific Revolutions*, Joe Jackson's *World on Fire*, and my own *Invention of Air*. Charles Babbage's attempt to build the first computer is chronicled in Doron Swade's *The Difference Engine*. The story of *Apollo 13* is told in Jim Lovell and Jeffrey Kluger's *Lost Moon*.

CHAPTER 2: LIQUID NETWORKS

On the importance of carbon and liquid water to the origins of life, I recommend several sources: a collection of essays, edited by J. William Schopf, entitled *Life's Origin*; Philip Ball's imaginative "biography" of water, *Life's Matrix*; and Carl Zimmer's *Science* essay "Evolutionary Roots: On the Origin of Life on Earth." The original Miller-Urey experiment was published in *Science* in the essay "A Production of Amino Acids Under Possible Primitive Earth Conditions." Silicon-based life appears in multiple science fictions, including Stanley Weinbaum's *A Martian Odyssey* and in the form of the Horta, a silicon-based creature discovered in episode 26 of the original *Star Trek* series. Chris Langton's theories about the generative power of liquid networks are developed in his essay "Life at the Edge of Chaos." Illuminating accounts of his work appear in both James Gleick's *Chaos* and Kevin Kelly's *Out of Control*. Wikipedia maintains an excellent "timeline of innovations," which provided a useful starting point for the charts of historical innovation that are included in this book. On the emergence and innovations of early Renaissance towns, Braudel's *Wheels of Commerce* remains the canonical text. The history of double-entry accounting is told in John Richard Edwards's *History of Financial Accounting*. For more on the power of collective decision-making, see James Surowiecki's *Wisdom of Crowds*, Howard Rheingold's *Smart Mobs*, Clay Shirky's *Here Comes Everybody*, and Kevin Kelly's *Out of Control*. Jaron Lanier's critique of the "hive mind" appears in his book *You Are Not a Gadget*, and in shorter form in the essay "Digital Maoism." For more on Kevin Dunbar's research, see "What Scientific Thinking Reveals About the Nature of Cognition." Malcolm Gladwell's take on the Jane Jacobsian future of workspace design appeared in the *New Yorker* in the essay "Designs for Working." Stewart Brand devotes a chapter of *How Buildings Learn* to the "low road" approach of Building 20. MIT also maintains a website that includes reminiscences about the building at http://libraries.mit.edu/archives/mithistory/building20/quotes.html.

CHAPTER 3: THE SLOW HUNCH

The intelligence failures surrounding the Phoenix Memo and the Moussaoui investigation are addressed in the *9/11 Commission Report* and in Bill Gertz's *Breakdown*. A transcript of Minneapolis field agent Coleen Rowley's letter to FBI director Mueller, detailing the failed connections leading up to the 9/11 attacks, is available at http://www.time.com/time/covers/1101020603/memo.html. The website http://www.historycommons.org/ contains an exhaustive archive of documents and press reports related to the 9/11 attacks, including the most comprehensive timeline of the late summer months preceding the attack that I have encountered. António Damásio's research into emotional brain flash assessments can be found in his artful work *Descartes' Error*. Snap judgments are also investigated in Gladwell's *Blink* and Jonah Lehrer's *How We Decide*. For more on Priestley's slow hunch, see my book *The Invention of Air*. Microsoft's principal scientist Bill Buxton writes about the slow hunch model in technology in his *BusinessWeek* essay "The Long Nose of Innovation." Howard Gruber's *Darwin on Man* is both the canonical study of Darwin's intellectual journey toward the idea of natural selection and one of the most insightful books on scientific creativity ever written. Images from Erasmus Darwin's commonplace book can be found online at http://www.revolutionaryplayers.org.uk/. John Mason's self-help guide to commonplace books appeared in his *Treatise on Knowledge*. Robert Darnton's essay "Extraordinary Commonplaces," from the *New York Review of Books*, provides an erudite account of the impact that commonplace books had on the Enlightenment-era literary imagination. Tim Berners-Lee's *Weaving the Web* tells the story of his invention of the Web, along with his ideas for improving the current platform. *Myths of Innovation* author Scott Berkun has an interesting analysis of Google's "innovation time off" program on his blog at http://www.scottberkun.com/blog/2008/thoughts-on-googles-20-time/.

CHAPTER 4: SERENDIPITY

For more on the battle between the chemical and electrical interpretations of brain activity, as well as additional material on Loewi's dream, see Eliot Valenstein's *The War of the Soups and the Sparks*. Edward O. Wilson's *Consilience* discusses the intellectual revelations of dreamwork, with specific reference to Kekulé's vision of Ouroboros. Ullrich Wagner's experiment is documented in the *Nature* essay "Sleep Inspires Insight." Robert Thatcher's study of different phase states can be found in "Intelligence and EEG Phase Reset" from the journal *NeuroImage*. For more on neurological serendipity, see David Robson's *New Scientist* essay "Disorderly Genius." William James's quote on the chaotic nature of "higher mind" appears in *Great Men, Great Thoughts, and the Environment.* For an entertaining and provocative overview of the evolution of sexual reproduction, see Matt Ridley's *Red Queen* and Jared Diamond's *Why Is Sex Fun?* John Barth's discussion of serendipity comes from his novel *The Last Voyage of Somebody the Sailor.* Henri Poincaré's pedestrian epiphanies are recounted in his *Foundations of Science.* A surprisingly long list of essays have argued that the Web is diminishing our opportunities for serendipitous discovery, including William McKeen's "The Endangered Joy of Serendipity" and Damon Darlin's "Serendipity, Lost in the Digital Deluge." Cass Sunstein has discussed his notion of an architecture of serendipity in *Going to Extremes*, and, with Richard Thaler, in *Nudge.* Alex Osborn's brainstorming technique was introduced in his book *Applied Imagination.* For a discussion of the problems with brainstorming and group creativity in general, see B. A. Nistad's "Illusion of Group Productivity," from the *European Journal of Social Psychology.* For more on open R&D labs, see Don Tapscott's *Wikinomics.*

CHAPTER 5: ERROR

For more on Lee de Forest's extraordinary career as an inventor (and, in later life, a Hollywood denizen) see his autobiography, *Father of Radio.* W. Rupert Maclaurin's essay "The Process of Technological Innovation" also contains revealing analysis of de Forest's error-prone invention of the triode. Additional information on Wilson Greatbatch's invention of the pacemaker can be found in John Adam's "Making Hearts Beat." Will Stanley Jevons's reference to the "errors of the great mind" appears in his "Principles of Science." For more on the generative power of error, see Kathryn Schulz's superb *Being Wrong.* I have discussed the connection between Kuhn's scientific paradigms and the long zoom approach in my *Invention of Air.* For an excellent discussion of Dunbar's research and the accidental discovery of cosmic background radiation, see Jonah Lehrer's *Wired* essay "Accept Defeat." A good introduction to Charlan Nemeth's research can be found in her essays "Differential Contributions of Majority and Minority Influence" and "Dissent as Driving Cognition, Attitudes, and Judgments." For a taste of the statistics of free association, see Palermo's *Word Association Norms.* A discussion of Darwin's failed theory of pangenesis can be found in Kirschner and Gerhart's *Plausibility of Life.* An overview of how the human genetic mutation rate was calculated can be found in Elie Dolgin's *Nature News* article "Human Mutation Rate Revealed." For more on

Susan Rosenberg's research on stress and mutation rates, see her essay "Microbiology and Evolution: Modulating Mutation Rates in the Wild" in *Science*. For more on the "fail fast" movement, see Doug Hall's *BusinessWeek* essay "Fail Fast, Fail Cheap" and Timothy Prestero's "Better by Design."

CHAPTER 6: EXAPTATION

Gutenberg's invention of the printing press is recounted in John Man's *Gutenberg*. I have also drawn upon the insights on Gutenberg's revolution that appear in Richard Ogle's *Smart World*, and Elizabeth Eisenstein's *Printing Press as an Agent of Change*. Gould and Vrba's concept of exaptation originally appeared in *Paleobiology* in the essay "Exaptation—A Missing Term in the Science of Form." For more on the concept, see Buss et al's *Adaptations, Exaptations, and Spandrels*. For more on the history of Google, see John Battelle's *The Search*. Franco Moretti discusses cultural exaptation in his essay "On Literary Evolution," included in his *Signs Taken for Wonders*. Koestler's *Act of Creation* contains many examples of exaptative thought, though he does not explicitly use the term, since the book predates Gould and Vrba's essay. For more on urban subcultures, see Claude Fischer's essays "Toward a Subcultural Theory of Urbanism" and "The Subcultural Theory of Urbanism: A Twentieth-Year Assessment." Jane Jacobs's *Death and Life of Great American Cities* and *The Economy of Cities* contain many similar insights about the capacity of big cities to cultivate small clusters of interests. (Chris Anderson discusses this in the context of his "long tail" theory in *The Long Tail*.) For more on the concept of the "Third Place," see Ray Oldenburg's *The Great Good Place*. For more on the innovations of the British coffeehouse, see Brian Cowan's *Social Life of Coffee*, Tom Standage's *History of the World in Six Glasses*, and my *Invention of Air*. Freud's Vienna salon is described in the context of innovation in Howard Gardner's *Creating Minds*. Martin Ruef's research appears in his essay "Strong Ties, Weak Ties and Islands," originally published in *Industrial and Corporate Change*. For more on Ronald Burt's analysis of social networks and organizational innovation, see his "Social Contagion and Innovation" and *Social Origins of Good Ideas*. Richard Ogle gives a riveting account of the exaptative creativity of Watson and Crick in *Smart World*. For more on Apple's design and development processes, see Lev Grossman's "How Apple Does It." Howard Gruber describes his "networks of enterprise" in his essay "The Evolving Systems Approach to Creative Work." For more on John Snow's diverse intellectual interests, see Peter Vinten-Johansen's *Cholera, Chloroform, and the Science of Medicine* and my *Ghost Map*.

CHAPTER 7: PLATFORMS

Charles Lyell's uniformitarian theory is outlined in his *Principles of Geology*. For more on Lyell's reaction to Darwin's idea, see the correspondence included in Darwin's *Autobiography*. For more on the concept of a keystone species, see R. T. Paine's "Conversation on Refining the Concept of Keystone Species," published in *Conservation Biology*. The concept of an ecosystem engineer is introduced in Clive Jones's "Organisms as Ecosystem Engineers." For a delightful history of the origins of GPS, see William Guier and George Weiffenbach's first-person account, "Genesis of Satellite Navigation." Franco Moretti has published a number of important works that look at the literary history of genres and devices, including *The Way of the World* and *Graphs, Maps, Trees*. For more on the innovation track record of the Twitter platform, see my essay "How Twitter Will Change the Way We Live" from *Time* magazine. Like many Web successes, Twitter's platform innovation relies on two key contributions from its users: end-user innovation and "venturesome" consumption. For more on these concepts, see Eric von Hippel's *Democratizing Innovation* and Amar Bhidé's *Venturesome Economy*. For more on the politics of collaborative platforms, see Clay Shirky's *Here Comes Everybody*. Tim O'Reilly discusses the idea of government-as-platform in a *Forbes* column titled "Gov 2.0: The Promise of Innovation." An account of the Redbird artificial reef can be found in Ian Urbina's *New York Times* article "Growing Pains for a Deep-Sea Home Built of Subway Cars." For more on the Jacobs vision of neighborhoods as emergent platforms, see my *Emergence*. Claudio Richter's coral research is described in John Roach's *National Geographic* article "Rich Coral Reefs in Nutrient-Poor Water: Paradox Explained?" For more on Calera's technology, see David Biello's *Scientific American* article "Cement from CO_2: A Concrete Cure for Global Warming?" I address the notion of the Web as rain forest in a *Discover* column, "Why the Web Is Like a Rainforest," and in a speech from the SXSW conference titled "Old Growth Media and the Future of News," a transcript of which is available at http://www.stevenberlinjohnson.com.

CONCLUSION: THE FOURTH QUADRANT

Willis Carrier's life story is told in his *Father of Air Conditioning*. Moretti's concept of "distant reading" is outlined in his *Graphs, Maps, Trees*. The innovation survey also draws on the histriometric approach to innovation developed by Dean Keith Simonton in *Genius, Creativity, and Leadership* and *Creativity in Science*. Yochai Benkler includes a more complex chart of potential innovation frameworks in *Wealth of Networks*. For more on the notion of collective invention, see Peter B. Meyer's "Episodes of Collective Invention." Marx and Engels's reaction to Darwin's work is discussed in Gruber's *Darwin on Man*. For more on the metaphor of the tangled bank and its importance to the theory of evolution, see Carl Zimmer's *Tangled Bank*. McPherson's dispute with Evans and his correspondence with Jefferson are discussed in Joseph Scott Miller's essay "Nonobviousness: Looking Back and Looking Ahead," included in the collection *Intellectual Property and Information Wealth*, edited by Peter K. Yu. I first encountered Jefferson's quote in Lawrence Lessig's *Future of Ideas*. That book, along with his books *Code* and *Remix*, is essential reading for anyone interested in the notion of an information commons.

Bibliography

Adam, John. "Making Hearts Beat." Lecture, Smithsonian Institution, Innovative Lives, 1999. http://invention.smithsonian.org/centerpieces/ilives/lecture09.html.

Anderson, Chris. *The Long Tail: Why the Future of Business Is Selling Less of More.* New York: Hyperion, 2008.

Baker, Jeffrey P. "The Incubator and the Medical Discovery of the Premature Infant." *Journal of Perinatology* 20, no. 5 (2000): 321–28.

Ball, Philip. *Life's Matrix: A Biography of Water.* New York: Farrar, Straus, and Giroux, 2000.

Barth, John. *The Last Voyage of Somebody the Sailor.* Boston: Little, Brown, 1991.

Bateson, Gregory. *Mind and Nature: A Necessary Unity.* Cresskill, N.J.: Hampton Press, 2002.

Battelle, John. *The Search: How Google and Its Rivals Rewrote the Rules of Business and Transformed Our Culture.* New York: Portfolio, 2006.

Baumol, William J. *The Free-Market Innovation Machine: Analyzing the Growth Miracle of Capitalism.* Princeton: Princeton University Press, 2002.

Benkler, Yochai. *The Wealth of Networks: How Social Production Transforms Markets and Freedom.* New Haven: Yale University Press, 2006.

Berkun, Scott. *The Myths of Innovation.* Sebastopol, Calif.: O'Reilly, 2007.

Berners-Lee, Tim. *Weaving the Web: The Original Design and Ultimate Destiny of the World Wide Web by Its Inventor.* New York: HarperCollins, 1999.

Bettencourt, L., J. Lobo, D. Helbing, C. Kühnert, and G. B. West. "Growth, Innovation, Scaling, and the Pace of Life in Cities." *Proceedings of the National Academy of Sciences* 104, no. 17 (2007): 7301.

Bhidé, Amar. *The Venturesome Economy*. Princeton: Princeton University Press, 2008.

Biello, David. "Cement from CO_2: A Concrete Cure for Global Warming?" *Scientific American* (August 7, 2008). http://tinyurl.com/ygp4o82.

Brand, Stewart. *How Buildings Learn: What Happens After They're Built.* New York: Penguin Books, 1994.

Braudel, Fernand. *A History of Civilizations*. New York: Penguin Books, 1993.

————. *The Perspective of the World*. Berkeley: University of California Press, 1992. Vol. 3, *Civilization and Capitalism 15th–18th Century*, 1992.

————. *The Wheels of Commerce*. Berkeley: University of California Press, 1992. Vol. 2, *Civilization and Capitalism 15th–18th Century*, 1992.

Brinkley, Joel. *Defining Vision: The Battle for the Future of Television*. New York: Harcourt Brace, 1997.

Brown, John Seely, and Paul Duguid. *The Social Life of Information*. Boston: Harvard Business School Press, 2002.

Burt, Ronald S. "Social Contagion and Innovation: Cohesion Versus Structural Equivalence." *American Journal of Sociology* 92, no. 6 (1987): 1287.

————. *Social Origins of Good Ideas*. Chicago: University of Chicago, 2003. http://www.uchicago.edu/fac/ronald.burt/research/SOGI.pdf.

Buss, David M., et al. "Adaptations, Exaptations, and Spandrels." *American Psychologist* 53, no. 5 (1998): 533–48.

Buxton, Bill. "The Long Nose of Innovation." *BusinessWeek* (January 2, 2008). http://www.businessweek.com/innovate/content/jan2008/id2008012_297369.htm.

Carrier, Willis Haviland. *Father of Air Conditioning*. Garden City, N.Y.: Country Life Press, 1952.

Cloud, John. "The Gurus of YouTube." *Time* (December 16, 2006). http://www.time.com/time/magazine/article/0,9171,1570721,00.html.

Cowan, Brian William. *The Social Life of Coffee: The Emergence of the British Coffeehouse*. New Haven: Yale University Press, 2005.

Cowan, Robin, and Nicolas Jonard. "The Dynamics of Collective Invention." *Journal of Economic Behavior and Organization* 52, no. 4 (2003): 513–32.

Csikszentmihalyi, Mihaly. *Creativity: Flow and the Psychology of Discovery and Invention*. New York: HarperPerennial, 1997.

———. *Flow: The Psychology of Optimal Experience.* New York: HarperPerennial, 1991.

Dacey, John S., Kathleen Lennon, and Lisa B. Fiore. *Understanding Creativity: The Interplay of Biological, Psychological, and Social Factors.* San Francisco: Jossey-Bass, 1998.

Damásio, António R. *Descartes' Error: Emotion, Reason, and the Human Brain.* London: Penguin, 2005.

Darlin, Damon. "Serendipity, Lost in the Digital Deluge." *The New York Times,* August 1, 2009.

Darnton, Robert. "Extraordinary Commonplaces." *New York Review of Books* 47, no. 20 (2000): 82–87.

Darwin, Charles. *Voyage of the* Beagle. Mineola, N.Y.: Dover Publications, 2002.

Darwin, Charles, and Francis, Sir Darwin. *The Autobiography of Charles Darwin.* Amherst, N.Y.: Prometheus Books, 2000.

Darwin, Charles, and R. D. Keynes. *Charles Darwin's* Beagle *Diary.* Cambridge, UK: Cambridge University Press, 1988.

De Forest, Lee. *Father of Radio: The Autobiography of Lee de Forest.* Chicago: Wilcox & Follett, 1950.

De Landa, Manuel. *A Thousand Years of Nonlinear History.* New York: Zone Books, 1997.

Diamond, Jared M. *Why Is Sex Fun? The Evolution of Human Sexuality.* New York: HarperCollins, 1997.

Dolgin, Elie. "Human Mutation Rate Revealed." *Nature News* (August 27, 2009). doi:10.1038/news.2009.864.

Dunbar, Kevin. "How Scientists Build Models: InVivo Science as a Window on the Scientific Mind." In *Model-based Reasoning in Scientific Discovery*, edited by Lorenzo Magnani, Nancy J. Nersessian, and Paul Thagard, 89–98. New York: Plenum Press, 1999.

———. "How Scientists Think: On-Line Creativity and Conceptual Change in Science." In *Creative Thought: An Investigation of Conceptual Structures and Processes*, edited by Thomas B. Ward, Stephen M. Smith, and Jyotsna Vaid, 461–93. Washington, D.C.: American Psychological Association, 1997.

———. "What Scientific Thinking Reveals About the Nature of Cognition." In *Designing for Science: Implications from Everyday, Classroom, and Professional Settings*, edited by Kevin Crowley, Christian D. Schunn, and Takeshi Okada, 115–40. Mahwah, N.J.: Lawrence Erlbaum Associates, 2001.

Edwards, John Richard. *A History of Financial Accounting.* London: Routledge, 1989.

Eisenstein, Elizabeth L. *The Printing Press as an Agent of Change*. Cambridge, UK: Cambridge University Press, 1980.

Fischer, Claude S. "The Subcultural Theory of Urbanism: A Twentieth-Year Assessment." *American Journal of Sociology* 101, no. 3 (1995): 543–77.

———. "Toward a Subcultural Theory of Urbanism." *American Journal of Sociology* 80, no. 6 (1975): 1319–41.

Florida, Richard L. *Cities and the Creative Class*. New York: Routledge, 2005.

———. *The Rise of the Creative Class and How It's Transforming Work, Leisure, Community and Everyday Life*. New York: Basic Books, 2004.

Franklin, Benjamin. *The Life and Writings of Benjamin Franklin, Vol. 2*. Philadelphia: McCarty & Davis, 1834.

Gardner, Howard. *Creating Minds: An Anatomy of Creativity Seen Through the Lives of Freud, Einstein, Picasso, Stravinsky, Eliot, Graham, and Gandhi*. New York: Basic Books, 1993.

Gay, Peter. *The Enlightenment: An Interpretation*. New York: W. W. Norton, 1977.

———. *The Enlightenment: The Science of Freedom*. New York: W. W. Norton, 1977.

Gertz, Bill. *Breakdown: The Failure of American Intelligence to Defeat Global Terror*. New York: Plume, 2003.

Gladwell, Malcolm. *Blink: The Power of Thinking Without Thinking*. New York: Back Bay Books, 2007.

———. "Designs for Working." *The New Yorker* (December 11, 2000): 60–70.

Gleick, James. *Chaos: Making a New Science*. New York: Penguin Books, 1987.

Gould, Stephen J., and Elisabeth S. Vrba. "Exaptation—A Missing Term in the Science of Form." *Paleobiology* 8, no. 1 (January 1982): 4–15.

Grossman, Lev. "How Apple Does It." *Time* (October 16, 2005). http://www.time.com/time/magazine/article/0,9171,1118384,00.html.

Gruber, Howard E. "Networks of Enterprise in Creative Scientific Work." In *Psychology of Science: Contributions to Metascience*, edited by Barry Gholson et al., 246. New York: Cambridge University Press, 1989.

———. "The Evolving Systems Approach to Creative Work." *Creativity Research Journal* 1, no. 1 (1988): 27–51.

Gruber, Howard E., Charles Darwin, and Paul H. Barrett. *Darwin on Man: A Psychological Study of Scientific Creativity*. New York: Dutton, 1974.

Guier, William H., and George C. Weiffenbach. "Genesis of Satellite Navigation." *Johns Hopkins APL Technical Digest* 19, no. 1 (1998): 15.

Hall, Doug. "Fail Fast, Fail Cheap." *BusinessWeek* (June 25, 2007). http://www
.businessweek.com/magazine/content/07_26/b4040436.htm_.

Harkness, Deborah E. *The Jewel House: Elizabethan London and the Scientific Revolution.* New Haven: Yale University Press, 2007.

Hausman, Carl R., and Albert Rothenberg. *The Creativity Question.* Durham: Duke
University Press, 1983.

Hippel, Eric von. *Democratizing Innovation.* Cambridge, Mass.: MIT Press, 2005.

Iberall, Arthur S. "A Physics for Studies of Civilization." In *Self-Organizing Systems:
The Emergence of Order*, edited by Eugene F. Yates. New York: Plenum Press,
1987.

Jackson, Joe. *A World on Fire: A Heretic, an Aristocrat, and the Race to Discover Oxygen.* New York: Penguin, 2007.

Jacobs, Jane. *The Nature of Economies.* New York: Modern Library, 2000.

———. *The Economy of Cities.* New York: Random House, 1969.

———. *The Death and Life of Great American Cities.* New York: Random House,
1961.

James, William. *Great Men, Great Thoughts, and the Environment.* Boston: Houghton
Mifflin, 1880.

Jastrow, Joseph, ed. *Story of Human Error.* Manchester, N.H.: Ayer Publishing, 1936.

Jefferson, Thomas. *The Writings of Thomas Jefferson, Vol. 1–19.* Edited by Albert El-
lery Bergh. Washington, D.C.: Thomas Jefferson Memorial Association, 1905.

Jevons, Will Stanley. "Principles of Science." *Daedalus* 87, no. 4 (1958): 148–54.

Johnson, George. "Of Mice and Elephants: A Matter of Scale." *New York Times*, January 12, 1999.

Johnson, Steven. "How Twitter Will Change the Way We Live." *Time* (June 5,
2009). http://www.time.com/time/business/article/0,8599,1902604,00
.html#ixzz0mu8f4umZ.

Johnson, Steven. *The Invention of Air: A Story of Science, Faith, Revolution, and the
Birth of America.* New York: Riverhead Books, 2008.

———. *Everything Bad Is Good for You: How Today's Popular Culture Is Actually
Making Us Smarter.* New York: Riverhead Books, 2006.

———. "Why the Web Is Like a Rainforest." *Discover* (October 2005). http://dis
covermagazine.com/2005/oct/emerging-technology.

———. *Emergence: The Connected Lives of Ants, Brains, Cities, and Software.* New
York: Scribner, 2002.

————. *Interface Culture: How New Technology Transforms the Way We Create and Communicate.* San Francisco: HarperEdge, 1997.

Jones, Clive G., J. H. Lawton, and M. Shachak. "Organisms as Ecosystem Engineers." *Oikos* (1994): 373–86.

Kauffman, Stuart A. *Investigations.* New York: Oxford University Press, 2000.

————. *At Home in the Universe: The Search for the Laws of Self-Organization and Complexity.* New York: Oxford University Press, 1995.

Kelly, Kevin. *Out of Control.* New York: Addison-Wesley, 1994.

Kirschner, Marc, and John Gerhart. *The Plausibility of Life: Resolving Darwin's Dilemma.* New Haven: Yale University Press, 2005.

Koestler, Arthur. *The Act of Creation.* London: Hutchinson, 1969.

Kostof, Spiro. *The City Shaped: Urban Patterns and Meanings Through History.* Boston: Little, Brown, 1991.

Kuhn, Thomas S. *The Structure of Scientific Revolutions.* Chicago: University of Chicago Press, 1970.

Langton, Christopher G., et al. "Life at the Edge of Chaos." *Artificial Life II* 10 (1992): 41–91.

Lanier, Jaron. *You Are Not a Gadget: A Manifesto.* Waterville, Maine: Thorndike Press, 2010.

————. "Digital Maoism: The Hazards of the New Online Collectivism." *The Edge* 183 (May 30, 2006).

Lehrer, Jonah. *How We Decide.* Boston: Mariner Books, 2010.

————. "Accept Defeat: The Neuroscience of Screwing Up." *Wired* (December 21, 2009). http://www.wired.com/magazine/2009/12/fail_accept_defeat/all/1.

Lessig, Lawrence. *Remix: Making Art and Commerce Thrive in the Hybrid Economy.* New York: Penguin Press, 2008.

————. *The Future of Ideas: The Fate of the Commons in a Connected World.* New York: Random House, 2001.

————. *Code, and Other Laws of Cyberspace.* New York: Basic Books, 1999.

Lovell, Jim, and Jeffrey Kluger. *Lost Moon: The Perilous Voyage of* Apollo 13. Boston: Houghton Mifflin, 1994.

Lyell, Charles, Sir. *Principles of Geology: Being an Inquiry How Far the Former Changes of the Earth's Surface Are Referable to Causes Now in Operation.* London: J. Murray, 1835.

Maclaurin, W. Rupert. "The Process of Technological Innovation: The Launching of a New Scientific Industry." *American Economic Review* 40, no. 1 (1950): 90–112.

MacLeod, Christine, and Alessandro Nuvolari. "'The Ingenious Crowd': A Critical Prosopography of British Inventors, 1650–1850." *Journal of Economic History* 5 (2005): 66–85.

Man, John. *Gutenberg: How One Man Remade the World with Words*. New York: John Wiley & Sons, 2002.

Margulis, Lynn, and Dorion Sagan. *Microcosmos: Four Billion Years of Evolution from Our Microbial Ancestors*. Berkeley: University of California Press, 1997.

Mason, John. *A Treatise on Self Knowledge: Showing the Nature and Benefit of That Important Science, and the Way to Attain It*. Boston: J. Loring, 1833.

McKeen, William. "The Endangered Joy of Serendipity." *St. Petersburg* [Florida] *Times*, March 26, 2006.

Meyer, Peter B. "Episodes of Collective Invention." U.S. Bureau of Labor Statistics Working Paper WP-368 (2003).

Miller, Stanley L. "A Production of Amino Acids Under Possible Primitive Earth Conditions." *Journal of NIH Research* 5 (1993).

———. "Production of Some Organic Compounds Under Possible Primitive Earth Conditions." *Journal of the American Chemical Society* 77, no. 9 (1955): 2351–61.

Mills, L. Scott, Michael E. Soulé, and Daniel F. Doak. "The Keystone-Species Concept in Ecology and Conservation." *BioScience* 43, no. 4 (1993): 219–24.

Moretti, Franco. *Graphs, Maps, Trees: Abstract Models for a Literary History*. New York: Verso, 2005.

———. *Atlas of the European Novel, 1800–1900*. New York: Verso, 1998.

———. *The Way of the World: The Bildungsroman in European Culture*. Translated by Albert Sbragia. London: Verso, 1987.

———. *Signs Taken for Wonders: Essays in the Sociology of Literary Forms*. New York: Verso, 1983.

Mumford, Lewis. *The City in History: Its Origins, Its Transformations and Its Prospects*. New York: Harcourt, Brace, Jovanovich, 1961.

National Commission on Terrorist Attacks Upon the United States. *The 9/11 Commission Report: Final Report of the National Commission on Terrorist Attacks Upon the United States*. New York: W. W. Norton, 2004.

Nemeth, Charlan Jeanne. "Dissent as Driving Cognition, Attitudes, and Judgments." *Social Cognition* 13, no. 3 (1995): 273–91.

Nistad, B. A., Wolfgang Stroebe, Hein F. M. Lodewijkx. "The Illusion of Group Productivity: A Reduction of Failures Explanation." *European Journal of Social Psychology* 36, no. 1 (January/February 2006): 31–48.

————. "Differential Contributions of Majority and Minority Influence." *Psychological Review* 93, no. 1 (1986): 23–32.

Ogburn, William F., and Dorothy Thomas. "Are Inventions Inevitable? A Note on Social Evolution." *Political Science Quarterly* 37, no. 1 (1922): 83–98.

Ogle, Richard. *Smart World: Breakthrough Creativity and the New Science of Ideas.* Boston: Harvard Business School Press, 2007.

Oldenburg, Ray. *The Great Good Place: Cafés, Coffee Shops, Community Centers, Beauty Parlors, General Stores, Bars, Hangouts, and How They Get You Through the Day.* New York: Marlowe, 1997.

O'Reilly, Tim. "Gov 2.0: The Promise of Innovation." *Forbes* (August 10, 2009). http://www.forbes.com/2009/08/10/government-internet-software-technology-breakthroughs-oreilly.html.

Osborn, Alex Faickney. *Applied Imagination: Principles and Procedures of Creative Problem Solving.* New York: Scribner, 1963.

Paine, R. T. "Conversation on Refining the Concept of Keystone Species." *Conservation Biology* 9, no. 4(1995): 962–64.

Palermo, David Stuart, and James J. Jenkins. *Word Association Norms: Grade School Through College.* Minneapolis: University of Minnesota Press, 1964.

Pauhus, P. B., M. T. Dzindolet, G. Poletes, and L. M. Camacho. "Perception of Performance in Group Brainstorming: The Illusion of Group Productivity." *Personality and Social Psychology Bulletin* 19, no. 1 (1993): 78.

Paulus, Paul B., and Bernard Arjan Nijstad. *Group Creativity: Innovation Through Collaboration.* New York: Oxford University Press, 2003.

Poincaré, Henri, and George Bruce Halsted. *The Foundations of Science; Science and Hypothesis, the Value of Science, Science and Method.* New York: Science Press, 1921.

Powell, Walter W., and Stine Grodal. "Networks of Innovators." *Handbook of Innovation* (2005): 1009–31.

Prestero, Timothy. "Better by Design: How Empathy Can Lead to More Successful Technologies and Services for the Poor (Discussion of Design Case Narratives: Rickshaw Bank, Solar-Powered Tuki, FGN Pump)." *Innovations: Technology, Governance, Globalization* 5, no. 1 (2010): 79–93.

Priestley, Joseph. *Experiments and Observations on Different Kinds of Air and Other Branches of Natural Philosophy, Connected with the Subject . . . : Being the Former Six Volumes Abridged and Methodized, with Many Additions.* Birmingham, UK: Thomas Pearson, 1790.

Prigogine, Ilya, and Gregoire Nicolis. *Exploring Complexity*. New York: W. H. Freeman, 1989.

Rheingold, Howard. *Tools for Thought: The History and Future of Mind-Expanding Technology*. Cambridge, Mass.: MIT Press, 2000.

Ridley, Matt. *The Red Queen: Sex and the Evolution of Human Nature*. New York: HarperPerennial, 2003.

———. *Genome: The Autobiography of a Species in 23 Chapters*. New York: HarperCollins, 1999.

Roach, John. "Rich Coral Reefs in Nutrient-Poor Water: Paradox Explained?" *National Geographic* (November 7, 2001). http://news.nationalgeographic.com/news/2001/11/1107_keyholecoral.html.

Robson, David. "Disorderly Genius." *New Scientist* 2714 (2009): 34–37.

Rogers, Everett M. *Diffusion of Innovations*. New York: Free Press, 1983.

Rosenberg, Susan M., and P. J. Hastings. "Microbiology and Evolution: Modulating Mutation Rates in the Wild." *Science Signaling* 300, no. 5624 (2003): 1382.

Ruef, Martin. "Strong Ties, Weak Ties and Islands: Structural and Cultural Predictors of Organizational Innovation." *Industrial and Corporate Change* 11, no. 3 (2002): 427.

Sawyer, R. Keith. *Explaining Creativity: The Science of Human Innovation*. New York: Oxford University Press, 2006.

Schofield, Robert E., ed. *A Scientific Autobiography of Joseph Priestley (1733–1804): Selected Scientific Correspondence*. Cambridge, Mass.: MIT Press, 1966.

Schopf, J. William. *Life's Origin: The Beginnings of Biological Evolution*. Berkeley: University of California Press, 2002.

Schulz, Kathryn. *Being Wrong: Adventures in the Margin of Error*. New York: HarperCollins, 2010.

Shapin, Steven. *The Scientific Revolution*. Chicago: University of Chicago Press, 1996.

Shapin, Steven, and Simon Schaffer. *Leviathan and the Air-Pump: Hobbes, Boyle, and the Experimental Life*. Princeton: Princeton University Press, 1985.

Shirky, Clay. *Here Comes Everybody: The Power of Organizing Without Organizations*. New York: Penguin Press, 2008.

Simonton, Dean Keith. *Creativity in Science: Chance, Logic, Genius, and Zeitgeist*. Cambridge, UK: Cambridge University Press, 2004.

———. *Origins of Genius: Darwinian Perspectives on Creativity*. New York: Oxford University Press, 1999.

———. *Greatness: Who Makes History and Why*. New York: Guilford, 1994.

————. *Genius, Creativity, and Leadership: Historiometric Inquiries.* Cambridge, Mass.: Harvard University Press, 1984.

Standage, Tom. *A History of the World in Six Glasses.* New York: Walker, 2005.

Sternberg, Robert J. *Handbook of Creativity.* Cambridge, UK: Cambridge University Press, 1999.

Sunstein, Cass R. *Going to Extremes: How Like Minds Unite and Divide.* Oxford, UK: Oxford University Press, 2009.

Surowiecki, James. *The Wisdom of Crowds.* New York: Anchor, 2005.

Swade, Doron, and Charles Babbage. *The Difference Engine: Charles Babbage and the Quest to Build the First Computer.* New York: Viking, 2001.

Tapscott, Don, and Anthony D. Williams. *Wikinomics: How Mass Collaboration Changes Everything.* New York: Portfolio, 2008.

Thaler, Richard H., and Cass R. Sunstein. *Nudge: Improving Decisions About Health, Wealth, and Happiness.* New York: Penguin Books, 2009.

Thatcher, Robert W., D. M. North, and C. J. Biver. "Intelligence and EEG Phase Reset: A Two Compartmental Model of Phase Shift and Lock." *NeuroImage* 42, no. 4 (2008): 1639–53.

Urbina, Ian. "Growing Pains for a Deep-Sea Home Built of Subway Cars." *New York Times*, April 8, 2008. http://www.nytimes.com/2008/04/08/us/08reef.html.

Valenstein, Elliot S. *The War of the Soups and the Sparks: The Discovery of Neuro-transmitters and the Dispute over How Nerves Communicate.* New York: Columbia University Press, 2005.

Vinten-Johansen, Peter, et al. *Cholera, Chloroform, and the Science of Medicine: A Life of John Snow.* New York: Oxford University Press, 2003.

Wagner, Ullrich, Steffen Gais, Hilde Haider, Rolf Verleger, and Jan Born. "Sleep Inspires Insight." *Nature* 427, no. 6972 (2004): 352–55.

Waldrop, Mitchell M. *Complexity: The Emerging Science at the Edge of Order and Chaos.* New York: Simon and Schuster, 1992.

Weinbaum, Stanley Grauman. *A Martian Odyssey, and Others.* Reading, Pa.: Fantasy Press, 1949.

Wilson, Edward O. *Consilience: The Unity of Knowledge.* New York: Knopf, 1998.

Wright, Robert. *NonZero: The Logic of Human Destiny.* New York: Pantheon Books, 2000.

Yu, Peter K. *Intellectual Property and Information Wealth: Issues and Practices in the Digital Age.* Westport, Conn.: Praeger Publishers, 2007.

Zimmer, Carl. *The Tangled Bank: An Introduction to Evolution.* Greenwood Village, Colo.: Roberts & Co. Publishers, 2010.

———. "Evolutionary Roots: On the Origin of Life on Earth." *Science* 323, no. 5911 (2009): 198.

Zittrain, Jonathan L. *The Future of the Internet—And How to Stop It.* New Haven: Yale University Press, 2009.

Index

Absolute zero, 272

Accounting, double-entry, 56–57, 228, 251

Adjacent possible, 25–42, 48, 72, 107, 110, 133, 156, 183, 191, 231
   for Analytical Engine, 36–40, 133
   of biosphere, 32–33, 36, 156
   Darwin's Paradox and, 33, 245
   error and, 138
   of evolution, 32, 142, 147
   exploring, 58, 65, 102, 141
   of jazz, 192
   networks and, 45–47
   in origins of life, 51–52
   serendipity and, 109
   of Web, 34, 39

Adobe Flash, 14, 40, 189

AdSense, 93

Afghanistan, war in, 94

Air conditioning, 214–18, 220, 231, 280

Airplanes, 280

Algorithms
   computer, 94, 114, 116, 158, 270
   evolutionary, 79, 238

Al Qaeda, 70

Altmann, Richard, 278

Alvarez, Luis and Walter, 293

Amazon, 243

American Airlines Flight 77, 70

Analytical Engine, 36–40, 133, 157, 270

Analytic geometry, 258

Ansoff, Vincent, 287

Apollo 13 mission, 41–42

Apple, 124, 169–71, 200, 244, 291
   iPhone, 39, 169–70, 193, 195, 244
   iPod, 170
   Macintosh, 15

Application programming interfaces (APIs), 194, 209

Apps for America, 195–96

Apps for Democracy, 195–97

Archaea, 293

Archaeopteryx, 154, 156, 157, 174

Archimedes, 58, 110, 240

Aristotle, 255, 258

ARPANET, 221

Ashbery, John, 122

Aspirin, 272

Asteroid impact, extinction caused by, 293

Assembly lines, 283

Assmann, Richard, 280

Atomic reactors, 286

Atomic theory, 267

Atta, Mohamed, 75
Audion, 133–35, 157, 285
Automobiles, 28, 213, 277, 283
Avery-MacLeod-McCarthy experiment, 287
Avogadro, Count Amedeo, 267

Babbage, Charles, 36–40, 42, 135, 157, 270, 287
Bacon, Francis, 84
Balance-spring watches, 259–60
Ball bearings, 252
Barbier, Charles, 269
Bardeen, John, 288
Barometers, 258, 259
Barth, John, 108
Battelle, John, 122
Bayliss, William Maddock, 281
Beagle (ship), 4–7, 82, 177, 180
Beatles, the, 164
Becquerel, Henri, 279
Begun, Semi Joseph, 286
Behaim, Martin, 252
Bell, Alexander Graham, 275
Bell, John, 85
Bell Labs, 288, 290
Beneden, Edouard van, 276
Benkler, Yochai, 220*n*
Benz, Karl, 277
Benzene, structure of, 103, 109–10, 160, 275
Bernardos, Nikolai, 277
Berners-Lee, Tim, 88–90, 92–94, 117, 158, 188, 206, 218, 220–21, 294
Berti, Gasparo, 258
Bessemer, Henry, 272
Bharat, Krishna, 94–95
Bible, Gutenberg, 152, 153, 228
Bicycles, 268
Bifocals, 265
Big Bang, 139, 290
Bin Laden, Osama, 70, 94
Biosphere, adjacent possible of, 31–33, 36, 107, 136
Bīrūnī, Abū Rayhān al-, 266
Bishop, J. Michael, 292
BlackBerry, 193
Blood circulation, 257
Blood types, 279–80
Bohr, Niels, 283, 284

Bose Acoustics, 63
Bouchon, Basile, 268
Boyle, Robert, 259, 260
Brahe, Tycho, 254, 256
Braille, 269
Bramah, Joseph, 255
Brand, Stewart, 63
Brattain, Walter, 288
Braun, Karl Ferdinand, 279
Brazil, social networking sites in, 93
Brin, Sergey, 158
Britain, Battle of, 286
Brunelleschi, Filippo, 56
Buchner, Eduard, 276
Buffalo Forge Company, 214–15
Bunsen burner, 272–73
Burnell, Jocelyn Bell, 290
Burroughs, William Seward, 37, 277
Burt, Ronald, 166–67
Bush, George W., 94
Büyükkökten, Orkut, 93
Byrne, David, 164
Byron, Lord, 38, 270

Calculators, 37, 259, 277
Calculus, 258, 261
Calera, 205
California, University of
    Berkeley, 139, 160
    Davis, 8
    Santa Barbara, 203
    Santa Cruz, 203
Capitalism, 217, 220*n*, 236, 237
    agrarian, 244
    industrial, 230
    merchant, 56–57
Carbonated water, 264
Carnot, Sadi, 269
Carothers, Wallace, 286
Carrier, Willis Haviland, 214–18, 220, 231, 280
Carrier Engineering Corporation, 215–16
Cartwright, Edmund, 266
Cary Institute for Ecosystem Studies, 182
Cell division, 276
Cerf, Vincent, 291
CERN (European Laboratory for Particle Physics), 90–93, 294

Chadwick, James, 286
Chaos, 52, 62–63, 85, 103–6, 120, 148
Chardack, William, 136
Chen, Steve, 14, 16, 39, 189, 197
Chicago, University of, 49, 166, 288
China
    ancient, 255, 266
    Great Wall of, 203
    movable type invented in, 153, 251
    smallpox inoculation in, 266
Chloroform, 269–70
Chomsky, Noam, 63
Chromosomes, 282
Chronic Disease Institute (Buffalo,
        New York), 135
Chronometer, 264
Cities, 16–21, 59, 65, 115, 222, 245
    adjacent possible in, 33
    ancient, 53–54
    discarded spaces in, 199, 200
    Italian Renaissance, 54–58, 228, 235
    subcultures of, 160–62
    superlinear scaling of, 9–11, 75, 162
Clarkson, Martha, 64
Clausius, Rudolf, 269
Clinton, Bill, 196
Clocks, pendulum, 259
Collins, Wilkie, 159, 191
Colt, Samuel, 270
Columbia University, 35, 282
Comets, 254, 262
Complex numbers, 253
Computers, 39, 47, 75, 116, 270
    Analytical Engine as, 37–39
    early, 134, 157, 185, 188
    graphical interface with, 15–16, 160,
        291
    interconnection of, *see* Internet;
        World Wide Web
    mainframe, 217
    multitasking of, 172
    personal, 14, 163, 170, 224
    vacuum tubes and, 132–33
Concave lenses, 251
Constantz, Brent, 203–5
Contact lenses, 278
Continental drift, 283
Copernicus, Nicolaus, 226, 252–53, 256

Cordus, Valerius, 253
Cosmic rays, 283
Cotton gin, 266
CPU (central processing unit), 38, 270
Creative Commons, 125
Credé, Carl, 276
Cretaceous period, 154
Crick, Francis, 159, 168–68, 190, 289
Cristofori, Bartolomeo, 262
Csikszentmihalyi, Mihaly, 64
CT scan (computerized tomography), 291
Cubic equations, 253
Cumings, Alexander, 255
Cuneus of Leyden, 34
Curie, Marie and Pierre, 279

Daguerre, Louis, 134–35, 271
Daimler, Gottlieb, 277
Dalton, John, 267
Damadian, Raymond, 292
Damásio, António, 76
Daphnia (water flea), 107–8
Dark Ages, 151
Darlin, Damon, 118, 120
Darnton, Robert, 86–87
Darwin, Charles, 17, 19, 78–79, 89, 90, 245,
        269
    biblical argument against, 155
    on Keeling Islands, 3–7, 41, 177–81, 238
    Marx on, 236–37
    natural selection theory of, 79–82, 143,
        190, 273
    notebooks of, 83–86
    on origins of life, 29, 50
    side interests of, 171–72
    tangled bank metaphor of, 238–40
Darwin, Erasmus, 85–87
Darwin's Paradox, 5, 33, 181–82, 190, 202,
        245
Data.gov, 195
Davis, Miles, 191–92, 210
Dean, Howard, 197
Defense Department, U.S., 290, 291
De Forest, Lee, 131–35, 137, 157, 220, 231
Delaware Department of Natural Resources
        and Environmental Control,
        198–99
Dennett, Daniel, 190

Denucé, Jean-Louis-Paul, 276
Deoxyribonucleic acid, *see* DNA
Descartes, René, 258
Design that Matters, 27–28
DEVONthink, 114–16, 127
Dickens, Charles, 159, 191
Difference Engine, 36–38, 40
Digital Equipment Corporation, 63
Din, Taqi al-, 254
Dinosaurs, extinction of, 293
Diodorus Siculus, 240–41
Djerassi, Carl, 289
DNA, 20, 30, 52, 190, 234, 287, 290, 291
    complementary replication system of,
        159
    double-helix structure of, 289, 168
    forensic use of, 293
    natural selection and, 106–7, 142
    repair system in, 144–47
Doppler effect, 184–85
Dorian scale, 192, 210
Dorsey, Jack, 192–94
Double-entry accounting, 56–57, 228, 251
Drais, Karl von, 268
Dreams, 100–103, 105, 109, 116–17, 123,
        136, 275
Drosophila melanogaster, 282
Duchamp, Marcel, 165
Dujardin, Edouard, 158–59
Dunbar, Kevin, 59–62, 138
DuPont Corporation, 286
DVD players, 14
Dylan, Bob, 165, 191

Earthquakes, 276, 281
Eccles, John Carew, 101
Edison, Thomas Alva, 231, 276, 278
Einstein, Albert, 59, 219, 281, 284
EKG, 278
Electrical batteries, 34, 132, 267, 283
Electric motors, 269, 277
Electroencephalogram (EEG), 104
Electromagnetic spectrum, 131–32, 134
Electrons, 279, 283
Elevators, 272
Elizabeth I, Queen of England, 255
Elliptical orbits, 256
Encyclopaedia Britannica, 122–23

Endorphins, 292
Engelbart, Doug, 15, 160, 290
Engels, Friedrich, 236–38
England, 244, 254
    commons of, 244
    Enlightenment in, 84, 162, 188
    Industrial Revolution in, 230, 264
    Victorian, 87–88
ENIAC, 134, 157
Enlightenment, 84, 86, 162, 226, 228, 234
Eno, Brian, 163–65
Enquire software application, 88–90, 93
Enzymes, 275–76, 291
Erdapfel, 252
Error, 131–48, 156
    inventions generated by, 132–36
    noise and, 138–43, 145–48
    paradigm shifts and, 137–38
Ether, 172, 253–54, 270
Evans, Oliver, 240–42, 270
Evolution, 20, 29, 106, 142, 154–58
    adjacent possible in, 32, 142, 147
    Darwin's theory of, 78–80, 171–72, 269
    of facial expressions, 116
    mutation in, 34, 142–45
    see also Natural selection
Exaptation, 151–74, 192, 232, 244
    in coffeehouse model, 165–70
    in evolutionary theory, 153–57, 171–72
    in subcultures, 160–63
    in shared media, 163–65
Exposition Universelle (Paris, 1855), 37
Eyeglasses
    bifocal, 265
    concave lens, 251

Fabricius, Johannes and David, 257
Facebook, 118, 195, 196
Fahrenheit, Daniel Gabriel, 263
Falcon, Jean, 268
Falling bodies, law of, 258
Faraday, Michael, 269
Farnsworth, Philo, 285
Federal Bureau of Investigation (FBI),
        69–74, 94, 126–27
    Automated Case Support system, 69, 76,
        91, 92, 95, 127
    Counterterrorism, 72

Radical Fundamentalist Unit (RFU), 71, 72, 92, 127
Federal Communications Commission (FCC), 12*n*, 13
Ferdinand III, Holy Roman Emperor, 259
Fermi, Enrico, 287
Ferraris, Galileo, 277
Ferro, Scipione del, 253
Fick, Adolf Eugen, 278
Finley, James, 267
Fiore, Antonio, 253
Firearms, 256, 270, 273
Fischer, Claude, 160–62, 200
Fisher, Alva John, 282
Fitch, John, 266
FitzRoy, Vice Admiral James, 5, 177, 181
Flaubert, Gustave, 191
Fleming, Alexander, 134, 285
Flemming, Walther, 276
Fletcher, William, 280
Flintlock firing mechanism, 256, 270
Flush toilets, 255
Flying shuttle, 263
Ford Motor Company, 283
Forensics, DNA, 293
Foursquare, 207
France, medical establishment in, 25–26
Franklin, Benjamin, 148, 172, 228, 264, 265
Franklin, Rosalind, 168
Frasca, David, 71, 72, 91
Fraunhofer, Joseph von, 268
Freud, Sigmund, 101–2, 109, 162, 188
Fuchsian functions, 110–11
Fulton, Robert, 265–66

GABA, 101
Galápagos Islands, 82, 83
Galileo Galilei, 226, 234, 255–60, 263, 266
Galvani, Luigi, 267
Gamma ray bursts, 294
Gates, Bill, 112–13
Gatling gun, 273
General Electric, 134
Geological uniformitarianism, 269
Geometry, 111, 258
Gerhardt, Charles, 272
Germany
    technology companies in, 286

viticulture in, 152, 153, 167
    in World War I, 7
Germ theory, 274
Getting, Ivan, 290
Gilbert, William, 255
Ginsburg, Charles Paulson, 289
Gladwell, Malcolm, 62, 76, 167
Global Positioning System (GPS), 14, 186–87, 207, 209, 290
Global warming, 122, 293
Goddard, Robert H., 284
Godfrey, Thomas, 263
Goethe, Johann Wolfgang von, 56
Golgi, Camillo, 282
Goodyear, Charles, 271
Google, 17*n*, 91, 93–95, 113, 121–23, 127, 128, 158, 196, 200, 209, 243
    Alerts, 208
    Books, 114
    Gmail, 93
    Innovation Time Off, 93–94
    Maps, 194
    News, 94–95
    Search Products and User Experience, 94
Gore, Al, 12*n*, 192
Gould, Steven Jay, 29, 126, 153–54, 156, 239
Grand Alliance, 12*n*
Granovetter, Mark, 167
Graphical user interface (GUI), 15–16, 160, 291
Gravity, 262, 284
Gray, Elisha, 275
Great Barrier Reef, 203, 205
Greatbatch, Wilson, 135–36
Greeks, ancient, 151, 192
    mythology of, 103, 160
GreenXchange, 125, 126, 158
Gruber, Howard, 79–81, 172
Guericke, Otto von, 259
Guier, William, 183–88, 209, 290
Gunter, Edmund, 258
Gutenberg, Johannes, 152–53, 167–68, 218, 220, 221, 226, 251
Guthrie, Samuel, 270

Hadley, John, 263
Halley, Edmund, 262

Hancock, Thomas, 271
Hanjour, Hani, 70
Hargreaves, James, 264
Harrington, John, 255
Harriot, Thomas, 257
Harrison, Thomas, 264
Harvard University, 234, 288
Harvey, William, 257
Hayek, Friedrich von, 217
Hegelian dialectics, 237
Heisenberg, Werner, 284
Helicopters, 284
Heliocentric theory, 252–53
Henlein, Peter, 252
Heredity, 275, 282
Herschel, William and Caroline, 266
Hertz, Heinrich Rudolf, 131–32, 279
Hess, Victor, 283
Hewish, Antony, 290
Hewlett-Packard, 200
High-definition television (HDTV), 12–16,
    190
Highs, Thomas, 264
High-Z Supernova team, 294
Hippocrates, 272
Homebrew Computer Club, 163, 188
H1N1 virus, 33
Hooke, Robert, 228, 259–62
Hooker, Joseph, 50
Hopps, John, 288
Hormones, 281, 289
Hot air balloons, 266
Hounsfield, Godfrey, 291
Howard, Ron, 41
Howe, Elias, 271
HTML, 158, 189
HTTP, 189, 209
Hubble, Edwin, 285
Hughes, John, 292
Hunches, 69–95, 99, 101, 102, 246
    APL platform and, 185, 188
    Darwin's, 4, 50, 78–87, 177
    instant, 76–77
    about 9/11 terrorists, 69–76
    serendipity and, 109, 111, 113–15, 123–24,
        126–28, 244
    slow, 77, 78, 81, 83, 89–94, 103, 115,
        132–36, 204–5, 232

Hurley, Chad, 14, 16, 39, 189, 197
Huxley, T. H., 79
Huygens, Christian, 259

IBM, 125, 189
Incubators, 25–29, 42, 276
India, 70
    social networking sites in, 93
Indian Ocean tsunami (2004), 27
Induction motors, 277
Industrial Revolution, 230, 234, 265
Infant mortality, 25–27
Information spillover, 53, 58, 65, 162, 235,
    242
Ingenhousz, Jan, 265
"Intelligent design," 155
Internet, 18, 119, 189, 217, 221, 244, 291
    commercial value of, 187
    filters on, 120
    generative platform of, 243
    open architecture of, 220n
    start-ups on, 200
    transmission costs of idea sharing on,
        235
    see also World Wide Web
IQ scores, 104–5
Iran, political protests in, 192
IRC messaging platform, 193
Islam, 57, 251, 253
    fundamentalist, 70, 74, 91
Italy, Renaissance, 54, 56–58, 228, 251,
    253
Ive, Jonathan, 169, 170

Jacob, François, 29
Jacobs, Jane, 11, 161–62, 199–200
Jacquard, Joseph-Marie, 157, 268
James, William, 105–6, 138
Janssen, Zacharias and Hans, 255
Japan, 286
    earthquakes in, 276
    HDTV in, 12n
Javascript, 189
Jefferson, Thomas, 228, 240–42
Jeffreys, Alex, 293
Jenner, Edward, 266
Jet engines, 285–86
Jevons, William Stanley, 136–37

Jobs, Steve, 169, 170, 292
Johns Hopkins University, 291
    Applied Physics Laboratory, 183–85,
        187–88
Johnson, George, 9
Jones, Clive, 182
Jones, Greg, 74
Joyce, James, 59, 159, 191
Jupiter, moons of, 256

Karim, Jawed, 14, 16, 39, 189, 197
Kauffman, Stuart, 30–31, 32, 52, 107
Kay, Alan, 160
Kay, John, 263
Keeling Islands, 3–7, 41, 82, 177–81, 200,
    238
Kekulé von Stradonitz, Friedrich August,
        103, 109–10, 160, 275
Kelly, T. J., 291
Kelvin scale, 272
Kepler, Johannes, 159, 256, 257
Khosla, Vinod, 205
Kindle, 114
Kleiber, Max, 7–10
Koestler, Arthur, 58–59, 159
Kolhörster, Werner, 283
Korean Air Lines Flight 007, 186
Korschinsky, S., 34
Kosterlitz, Hans, 292
Krebs cycle, 287
Kuhn, Thomas, 59, 137
Kühne, Wilhelm, 275
Kundra, Vivek, 195, 197

Laennec, René, 268
Lamarckianism, 143
Landsteiner, Karl, 279
Langton, Christopher, 52
Lanier, Jaron, 58
Lasers, 290
Lavoisier, Antoine, 228, 265, 267
Lee, William, 255
Leeuwenhoek, Antonie Philips van, 255,
    260
Lehrer, Jonah, 139
Leibniz, Gottfried Wilhelm, 261
Lenormand, Louis-Sébastien, 252
Leonardo da Vinci, 56, 226, 252, 278

Lessig, Lawrence, 236, 242
Libby, Willard, 288
Liebig, Justus von, 270
Life, origins of, 29–32, 49–52
    simulation of, 49, 289
Light
    spectrum of, 260, 268
    speed of, 260–61, 281
Lightbulbs, 231, 276
Lightning rods, 264
Lilienthal, Otto, 280
Linnaeus, Carl, 228, 263
Lion, Alexandre, 26
Lippershey, Hans, 255
Liquid networks, 45–65, 75, 162, 169, 232,
    235, 246
Lithography, 267
Lloyd, Edward, 188–89
Locke, John, 84–85, 87, 228
Locomotives, 268
Loewi, Otto, 100–102
Logarithms, 8, 257–58
London, 163
    cholera in, 114–15, 172–74
    Science Museum, 37
    University, 168
    World's Fair (1862), 274
Long-zoom perspective, 20–21
Looms, mechanization of, 38, 85, 157, 263,
        266, 278
Loschmidt, Joseph, 275
Louis XIII, King of France, 256
Lovelace, Ada, 38, 39, 270
Lyell, Charles, 6, 82, 177, 179, 269

Magnetism, 255
Malthus, Thomas, 79–81, 83, 222
Maps
    Google, 194
    Mercator projection, 254
Marconi, Guglielmo, 131–32, 279
Marin le Bourgeoys, 256
Mariotte, Edme, 260
Marius, Simon, 256
Martin, Odile, 25, 26
Marx, Karl, 236–37
Mason, John, 86
Mason jars, 272

Massachusetts Institute of Technology (MIT), 27
  Building 20, 62–63, 65
Mass production, 152, 153, 230, 283
Mathematical symbols, 253
Mauna Kea, 178
Mauritius, 177
Maybach, Wilhelm, 277
Mayer, Marissa, 94
McClure, Frank T., 185–87, 209
McGaffey, Ives W., 274
McKeen, William, 117, 118, 120
McPherson, Isaac, 240, 242
Mellotron synthesizer, 164
Mendel, Gregor, 275
Mendeleev, Dmitri, 101, 274
Mendelian genetics, 190
Mercator projection, 254
Mesopotamia, 245
Metabolism, relationship of size to, 8–10
Michelangelo, 56
Microorganisms, 260
Microscopes, 61, 255, 260
Microsoft, 63, 112–13
  Building 99, 63–64
  Windows, 16, 244
  Windows Media Player, 14
Microwave ovens, 287
Microwaves, 183, 185, 290
Milky Way, 266
Miller, Stanley L., 49, 50, 289
Milne, John, 276
Milton, John, 59, 84
Mitochondria, 32, 278
Molecules, formulation of theory of, 267
Montgolfier, Joseph-Michel and Jacques-Étienne, 266
Moretti, Franco, 158, 222, 224
Morgan, Thomas Hunt, 275, 282
Morse code, 132, 270
Motion, laws of, 262
Motion picture camera, 278
Motorcycle, 277
Mount Vesuvius, 151
Moussaoui, Zacarias, 73–76, 91, 94, 127
MRI (magnetic resonance imaging), 292
Mueller, Robert, 71, 73
Muller, F. A., 278

Multiples, 34–35
Mutation, 34, 142, 144–47, 154–56, 282
Muybridge, Eadweard, 278

Nairobi (Kenya), 29, 126, 156, 239
Napier, John, 257–59
National Science Foundation (NSF), 203
Natural selection, 79–82, 106–7, 142–45, 155–56, 190, 273
Navigational devices, 37, 257, 263
  see also GPS
Navy, U.S., 135
Nemeth, Charlan, 139–42, 147
NeoNurture, 28
Neurotransmitters, 99, 282
Neutrons, 286, 287
Newcomen, Thomas, 262
Newlands, John, 274
Newton, Isaac, 228, 257, 258, 260–63, 286
New York City, 163, 200, 213–15
  Police Department (NYPD), 77
Nicholas of Cusa, 251
Niepce, Joseph Nicephore, 271
Nike Corporation, 125, 158
Nitroglycerine, 271
Nobel, Alfred, 231, 274
Nobel Prize, 101, 139, 169, 283, 290, 293
Noise, 104–5, 116, 120, 222, 232
  error and, 138–43, 145–48
Non-Euclidian geometry, 111
Nucleotides, 32, 52, 144, 289
Nylon, 286

Obama, Barack, 195
Ocean tides, 257
Octants, 263
Ogburn, William, 35
Ogle, Richard, 167–69
Ohain, Hans von, 285
Oldenburg, Ray, 162
Oldham, Richard, 281
Olszewski, Stanislav, 277
Oncogenes, 292
Onnes, Heike Kamerlingh, 282
Oral contraceptives, 234, 281, 288–89
Orbits
  of comets, 262
  of electron around nucleus of atom, 283

of man-made satellites, 183, 185–86, 290

planetary, 252, 256

O'Reilly, Tim, 197

Orkut, 93

Osborn, Alex, 126

Otis, Elisha, 272

Otto, Nikolaus, 277

Oughtred, William, 257–58

Outside.in, 208, 209

Oxford University, 84

Oxygen, isolation of, 34, 35, 265

Ozzie, Ray, 112–13

Pacemakers, 136, 288

Pacioli, Luca, 57, 251

Page, Larry, 158

PageRanks, 158

Pan Am International Flight Academy, 73, 74

Pangea, 283

Papin, Denis, 261, 262

Paracelsus, 254

Parachutes, 252

Paradigm shifts, 137–38

Paris, 163

Maternité de, 25–26

University of, 253

Parkes, Alexander, 274

Pascal, Blaise, 259

Pasteur, Louis, 274, 276

Pelouze, J. T., 271

Pencils, 254

Pendulums, 259, 276

Penicillin, 134, 285

Penzias, Arno, 139, 290

Periodic table, 50, 101, 274

Perkins, Jacob, 270

Pescara, Raúl Pateras, 284

Phoenix memo, 69–73, 75–76, 91, 94, 127

Photography, 35, 135, 271

Photosynthesis, 265

Piano, 262

Pi Sheng, 153, 167

Planck, Max, 284

Plante, Gaston, 272

Plant respiration, 78, 265

Plastic, 33, 48, 156, 274

Platforms, 21, 40, 93, 125, 177–210, 234

city as, 245

emergent, 187–88, 194, 196, 197, 232, 234

generative, 189, 243–44

open, 125, 178, 194, 198, 210, 221, 235, 243

stacked, 189–94, 206, 209, 210

Pliny the Elder, 151

Poe, Edgar Allan, 191

Poincaré, Henri, 110–13, 245

Poindexter, Admiral John, 73

Polaris nuclear missiles, 186, 209

Portland cement, 204

Pressure cookers, 261

Prestero, Timothy, 27–28

Priestley, Joseph, 34, 35, 77–78, 86, 138, 172, 218, 228, 264, 265

Princeton University, 139, 234

Printing press, 152–53, 221, 226, 251

Procter & Gamble, 125

Proust, Joseph, 267

Ptolemaic astronomy, 152

Public Broadcasting System (PBS), 122

Public Enemy, 165

Pulmonary respiration, 253

Pulsars, 290

Punch cards, 38, 157, 268

Pyramids, 203

Quantum mechanics, 284

Quarter-power laws, 8–11

Quick-Time, 14

Radioactivity, 279

Radiocarbon dating, 288

Radios, 13, 35, 132–35, 164, 165, 224, 279

RAM (random access memory), 38

Ramón y Cajal, Santiago, 282

Rangiroa atoll, 204

Raytheon Corporation, 166–67, 290

Reagan, Ronald, 186

Red Sea, 201

Refrigerators, 270

Relativity theory, 219, 281, 284

REM sleep, 101, 102

Renaissance, 54, 56, 151, 226, 228, 234, 253

Reproductive strategies, 106–8, 146–47

Research and development (R&D) labs, 124–25, 127, 158, 233

Respiration
plant, 78, 265
pulmonary, 253

Restriction enzymes, 291

Revolvers, 270

Richter, Claudio, 201

Riess, Adam, 294

RNA, 290–93

Roberts, Richard J., 292–93

Rock, John, 288

Rockets, 284

Roemer, Olaus, 260–61

Roentgen, Wilhelm, 279

Romans, ancient, 151

Rosen, Jonathan, 28

Rosenberg, Susan, 145–46

Royal Air Force (RAF), 285

Royal Society, 228

Rubber, 126
vulcanized, 136, 271

Rudolff, Christoph, 253

Ruef, Martin, 165–67

Rumsey, James, 266

Rutherford, Ernest, 283

Sackett-Wilhelm Lithography Company, 213–15

Salesforce.com, 128

Samit, Harry, 74

Sanger, Margaret, 289

Santa Fe Institute, 9

Sarcopterygii, 156

Saudi Arabia, 70

Savery, Thomas, 262

Sawyer, William, 276

Schawlow, Arthur L., 290

Scheele, Carl Wilhelm, 34, 35, 265

Scheutz, Per Georg, 37

Schickard, Wilhelm, 259

Schmidt, Brian, 294

Schumpeter, Joseph, 217n

Scleractinia, 178, 182

Seismographs, 276

Senebier, Jean, 265

Senefelder, Alois, 267

September 11, 2001, terrorist attacks (9/11), 69–76, 91, 94, 95

Serendipity, 99–128, 156, 232, 246
chaos and, 103–6, 120
in dreams, 100–103, 105, 109, 116–17
hunches and, 109, 111, 113–15, 123–24, 126–28, 244
Web and, 117–23

Servetus, Michael, 253

Sewing machines, 280

Sexual reproduction, 106–8, 146–47

SGML, 189

Sharp, Philip A., 292–93

Shibh, Ramzi bin al-, 75

Shocklee, Hank, 165

Shockley, Bill, 288

Sholes, Christopher Latham, 275

Shore, John, 262

Singer, Isaac, 271

Six Sigma, 148

Sketchpad, 291

Slide rules, 257–58

Smallpox vaccine, 266

Smith, H. O., 291

SMS mobile communications platform, 192, 209

Snow, John, 172, 173, 218

Sobrero, Ascanio, 271

Solar system, heliocentric theory of, 252–53

Sony Corporation, 13

Soubeiran, Eugène, 270

Soubra, Zakaria Mustapha, 70

Soviet Union, 186, 209
space program of, 183–88

Spectroscopes, 268

Speed of light, 260–61, 281

Spencer, Percy, 288

Spillover, information, 53, 58, 65, 162, 235, 242

Spinning jenny, 264

Sputnik, 183–88

Stanford University, 234, 291
Business School, 165
Woods Institute for the Environment, 204

Staphylococcus, 134

Starling, Ernest Henry, 281

Steamboats, 265–66

Steam engines, 34–35, 230, 254, 262–63

Steam locomotives, 268
Steelmaking, 272
Stocking frames, 85, 255
Stone, Biz, 192–94
Strasburger, Eduard, 276
Stratosphere, 280
StumbleSafely, 195
Subcultures, 160–62, 200
Sunlight Foundation, 196
Sunspots, 34, 256–57
Sunstein, Cass, 119, 120
Superconductivity, 282
Superlinear scaling, 9–11, 75, 162
Supernovas, 254, 294
Suspension bridges, 267
Sutherland, Ivan, 291
Swan, Joseph, 276
Switzerland, 7
Syntex, 289
Szilard, Leo, 287

Talking Heads, 164
Tangled bank, metaphor of, 197, 200, 245,
    246
  Darwin's use of, 238, 239, 240
Tansley, Arthur, 287
Tape recorders, 164, 165, 286
Tarnier, Stéphane, 25–26, 276
Tartaglia, Niccolò, 253
Taxonomy, 263
TBWA/Chiat/Day advertising agency, 62
TCP/IP, 209, 221
Teisserenc de Bort, Léon, 280
Telegraphy, 34, 131, 270, 279
Telephone, 34, 133, 275
Telescopes, 256, 257, 263
Television, 11–15, 132, 133, 231, 285
10/10 rule, 13, 16
Terrestrial globes, 252
Tesla, Nikola, 132, 277, 279
Textile industry, 38, 255, 263, 266
Thatcher, Robert, 104–5
Thermodynamics, 269
Thermometers, 263
Thomas, Dorothy, 35
Thomson, J. J., 279
3M Corporation, 127
Tompion, Thomas, 260

Torricelli, Evangelista, 258, 259
Total Information Awareness project, 73
Total Quality Management, 148
Townes, Charles H., 290
Townsend, John, 279
Transistors, 135, 288
Transit system project, 186
Trevithick, Richard, 268
Tuning forks, 262
Turing, Alan, 287
Twitter, 118, 172, 192–94, 196, 197, 207–9,
    243

Uncertainty principle, 284–85
Uniform acceleration, 258
United Technologies, 216
UNIVAC, 185, 188
Universal gravitation, law of, 262
Universe, expansion of, 285, 294
Unwin, William, 188–89
Urey, Harold C., 49, 50, 289

Vacuum cleaners, 274
Vacuum pumps, 259
Vacuum tubes, 35, 39, 133–35, 157, 220, 231,
    285, 288
Vail, Alfred, 270
Varmus, Harold, 292
Vaughan, Philip, 252
VCRs (video cassette recorders), 11, 13–14,
    289
Venice, 57, 251
Vernier scale, 257
Victorian era, 87
  sewage systems of, 114–15
  technology of, 37
Vitamins, 280–81
Volta, Alessandro, 267
Von Kleist, Dean, 34
Vrba, Elisabeth, 153–54
Vries, Hugo de, 34
Vulcanized rubber, 136, 271

Wagner, Ullrich, 102
Wallace, Alfred Russel, 79, 273
Waller, Augustus, 278
Walpole, Horace, 108
Washing machines, 282

Washington, D.C., 195–97
Watches, 228, 252, 259–60
Watson, James, 167–68, 190, 289
Watson-Watt, Robert, 286
Watt, James, 230
Wegener, Alfred, 283
Weiffenbach, George, 183–88, 209, 290
Welding, 277
Welsbach burner, 132
Wendt, William F., 214
West, Geoffrey, 9–11
Whewell, William, 269
Whitney, Eli, 266
Whittle, Frank, 285
Wikipedia, 17n, 117, 122–23, 169
Wilcox, K. W., 291
Williams, Evan, 192–94
Williams, Ken, 69–76, 91, 92
Wilson, H. A., 279
Wilson, Robert Woodrow, 139, 290
Wirth, Louis, 160
Woese, Carl, 290, 293
World War I, 7
World War II, 38, 63, 135
World Wide Web, 16–18, 21, 46, 93, 114,
        148, 219, 221, 245
    adjacent possible of, 34, 39

development of, 89–90, 294
ecosystem of, 206–7
generative platform of, 189, 210, 243, 244
geographic mashups on, 194
hunch database on, 127
marketplaces on, 125
news on, 94
politics and, 197
queries of, 91 (see also Google)
serendipity and, 109, 117–23
superlinear scaling in, 75
video and sound on, 14–15 (see also
        YouTube)
work environments and, 62
see also Internet
Wozniak, Steve, 170, 292
Wright brothers, 280
Wulf, Theodor, 283

Xerox, 15, 16, 291
X-ray crystallography, 168
X-rays, 34, 291, 293

YouTube, 14–16, 34, 39–40, 189–90, 192, 244

Zuckerman, Ethan, 119
Zuse, Konrad, 287

Steven Johnson writes about big ideas.

The world is changing—faster than ever. There are more big ideas and more good ideas out there. And Steven Johnson tells us both where they came from and where they can take us.

His books are insightful, wide-ranging, about the future and about our history. They are essential for business, innovation, technology, history, and science readers.

Bill Clinton gave a talk recently where he discussed some of Steven Johnson's books:

"There's an interesting book—if you want to be optimistic about the future—by Steven Johnson, who's a great science writer. It's called *Future Perfect*. [Two of his earlier] books, one of them is called *The Ghost Map*, which is about how the cholera epidemic was solved in London; and one's called *The Invention of Air*, which is about the discovery of oxygen."

Steven Johnson's curious, dynamic, creative mind reveals a fascinating world of ideas and innovation.

Steven Johnson has big ideas.

T282-0413

Everything Bad Is Good for You
How Today's Popular Culture Is Actually Making Us Smarter

Steven Johnson's hallmark classic on pop culture and technology. In this provocative, unfailingly intelligent, thoroughly researched, and convincing book, Johnson draws from fields as diverse as neuroscience, economics, and media theory to argue that the pop culture we soak in every day—from *The Lord of the Rings* to Grand Theft Auto to *The Simpsons*—is actually sophisticated and, far from rotting our brains, is actually posing new cognitive challenges that are making our minds immeasurably sharper.

"Iconoclastic and captivating."

—The Boston Globe

"Persuasive…The old dogs won't be able to rest as easily once they've read *Everything Bad Is Good for You*, Steven Johnson's elegant polemic."

—Walter Kirn, *The New York Times Book Review*

"Wonderfully entertaining. Steven Johnson proposes that what is making us smarter is precisely what we thought was making us dumber: popular culture."

—Malcolm Gladwell, *The New Yorker*

The Ghost Map: The Story of London's Most Terrifying Epidemic—and How It Changed Science, Cities, and the Modern World

A *New York Times* Notable Book

A riveting page-turner about a real-life historical hero, Dr. John Snow. In the summer of 1854, London has just emerged as one of the first modern cities in the world. But lacking the infrastructure—garbage removal, clean water, sewers—necessary to support its rapidly expanding population, the city has become the perfect breeding ground for a terrifying disease no one knows how to cure. As the cholera outbreak takes hold, a physician and a local curate are spurred to action—and ultimately solve the most pressing medical riddle of their time.

Johnson illuminates the intertwined histories and interconnectedness of the spread of disease, contagion theory, the rise of cities, and the nature of scientific inquiry.

"Thrilling." **—GQ** "Vivid." **—The New Yorker**

"Marvelous." **—The Wall Street Journal** "Fascinating."
—The New York Times Book Review

The Invention of Air
A Story of Science, Faith, Revolution, and the Birth of America

A book of world-changing ideas wrapped around a compelling narrative, a story of genius and violence and friendship in the midst of sweeping historical change that provokes us to recast our understanding of the Founding Fathers. *The Invention of Air* is the story of Joseph Priestley—scientist and theologian, protégé of Benjamin Franklin, friend of Thomas Jefferson—an eighteenth-century radical thinker who played pivotal roles in the invention of ecosystem science, the discovery of oxygen, the founding of the Unitarian Church, and the intellectual development of the United States. Steven Johnson here uses a dramatic historical story to explore themes that have long engaged him: innovation and the way new ideas emerge and spread, and the environments that foster these breakthroughs.

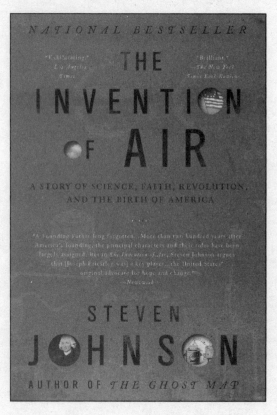

"Entertaining...clear-sighted and intelligent."

—The New Yorker

"An exemplar of the postcategorical age...Brilliant."

—The New York Times Book Review

The Innovator's Cookbook
Essentials for Inventing What Is Next

Innovation is one of today's buzzwords for a reason. The need to push forward, find new paths and new ideas in an ever-evolving world, is a vital part of business, of education, of politics, of our daily lives. This is an essential book for anyone interested in innovation: the key texts on the topic from a wide range of fields, as well as interviews with successful, real-world innovators, prefaced with an original essay from Johnson that draws upon his own experience as an entrepreneur and author.

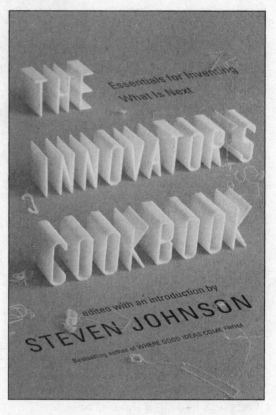

Includes writing from:

Stewart Brand

Clayton Christensen

Richard Florida

Teresa Amabile

Peter Drucker

Amar Bhidé

Ray Ozzie

Beth Noveck

Jon Schnur

Katie Salen

Brian Eno

Future Perfect
The Case for Progress in a Networked Age

Steven Johnson makes the case that a new model of political change is on the rise, transforming everything from local governments to classrooms, from protest movements to health care. Johnson paints a compelling portrait of this new political worldview—influenced by the success and interconnectedness of the Internet, but not necessarily dependent on high-tech solutions—that breaks with traditional categories of liberal or conservative thinking.

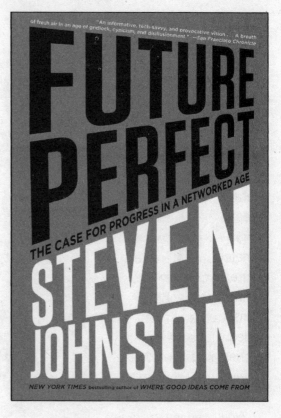

With his acclaimed gift for multidisciplinary storytelling and big ideas, Johnson explores this innovative vision of progress through a series of fascinating narratives. At a time when the conventional wisdom holds that the political system is hopelessly gridlocked with old ideas, *Future Perfect* makes the timely and uplifting case that progress is still possible, and that new solutions are emergent.

"An informative, tech-savvy, and provocative vision... A breath of fresh air in an age of gridlock, cynicism, and disillusionment."

—San Francisco Chronicle

"In clear and engaging prose, Johnson writes about this emerging movement...*Future Perfect* is a buoyant and hopeful book."

—The Boston Globe

T284-0413

How We Got to Now
Six Innovations That Made the Modern World

In this illustrated volume, Steven Johnson explores the history of innovation over centuries, tracing facets of modern life (refrigeration, clocks, and eyeglass lenses, to name a few) from their creation by scientists, engineers, hobbyists, amateurs, and entrepreneurs to their unintended historical consequences. Filled with surprising stories of accidental genius and brilliant mistakes, *How We Got to Now* investigates the secret history behind contemporary life.

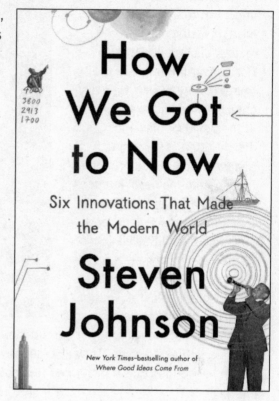

Accompanied by a major six-part television series on PBS, *How We Got to Now* is the story of collaborative networks building the modern world, written in the provocative, informative, and engaging style that has earned Johnson fans around the globe.

WONDERLAND
How Play Made the Modern World

Play has always been more important to innovation and creativity than most people realize. Just as *How We Got to Now* investigated the secret history behind everyday objects, this vivid examination of the power of play and delight offers a surprising history of popular entertainment. Roving from medieval kitchens and ancient taverns to casinos, theaters, computer labs, and shopping malls, Steven Johnson locates the cutting edge of innovation wherever people are working hardest to keep themselves and others amused.

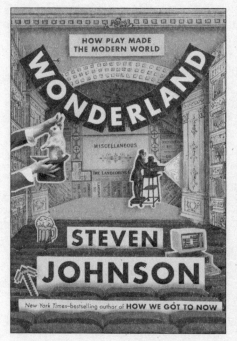

These wonderlands of amusement did more than just entertain their patrons, Johnson argues. They also directly contributed to economic and social revolutions that transformed the modern world.

"A house of wonders itself . . . *Wonderland* inspires grins and well-what-d'ya-knows."
—The New York Times Book Review

"A rare gem . . . Our illogical, enduring fascination with play remains one of life's great mysteries. That is precisely what makes the subject so fascinating, and *Wonderland* such a compelling read."
—The Washington Post

"Johnson's prose is nimble, his knowledge impressive. . . . *Wonderland* is original and fun, as well it should be, given the subject."
—San Francisco Chronicle